Routledge Annals of Bioethics

Series Editors:
Mark J. Cherry
St. Edward's University, USA
Ana Smith Iltis
Saint Louis University, USA

For a full list of titles in this series, please visit www.routledge.com

8 **Practical Autonomy and Bioethics**
James Stacey Taylor

9 **The Ethics of Abortion**
Women's Rights, Human Life, and the Question of Justice
Christopher Kaczor

10 **Bioethics, Public Moral Argument, and Social Responsibility**
Edited by Nancy M. P. King and Michael J. Hyde

11 **The Ethics of Gender-Specific Disease**
Mary Ann Cutter

12 **Death, Posthumous Harm, and Bioethics**
James Stacey Taylor

13 **Human Dignity in Bioethics**
From Worldviews to the Public Square
Edited by Stephen Dilley and Nathan J. Palpant

14 **Parental Obligations and Bioethics**
The Duties of a Creator
Bernard G. Prusak

15 **The Bioethics of Pain Management**
Beyond Opioids
Daniel S. Goldberg

16 **The Ethics of Pregnancy, Abortion, and Childbirth**
Exploring Moral Choices in Childbearing
Helen Watt

17 **Medical Ethics, Prediction, and Prognosis**
Interdisciplinary Perspectives
Edited by Mariacarla Gadebusch Bondio, Francesco Spöring, and John-Stewart Gordon

Medical Ethics, Prediction, and Prognosis

Interdisciplinary Perspectives

Edited by
Mariacarla Gadebusch Bondio,
Francesco Spöring,
and John-Stewart Gordon

Routledge
Taylor & Francis Group
NEW YORK AND LONDON

First published 2017
by Routledge
605 Third Avenue, New York, NY 10017

and by Routledge
4 Park Square, Milton Park, Abingdon, Oxon OX14 4RN

Routledge is an imprint of the Taylor & Francis Group, an informa business

Library of Congress Cataloging-in-Publication Data
Names: Gadebusch Bondio, Mariacarla, editor.
Title: Medical ethics, prediction, and prognosis : interdisciplinary perspectives / edited by Mariacarla Gadebusch Bondio, Francesco Spöring, and John-Stewart Gordon.
Description: 1 [edition]. | New York : Routledge, 2017. | Series: Routledge annals of bioethics ; 17 | Includes bibliographical references and index.
Identifiers: LCCN 2016056798 | ISBN 9781138632691 (hardback : alk. paper)
Subjects: LCSH: Medical ethics. | Medicine—Forecasting.
Classification: LCC R724 .M29426 2017 | DDC 174.2—dc23
LC record available at https://lccn.loc.gov/2016056798

ISBN: 978-1-138-63269-1 (hbk)
ISBN: 978-1-315-20808-4 (ebk)

Typeset in Sabon
by Apex CoVantage, LLC

Contents

Introduction

Predictive Medicine—An Interdisciplinary Approach

Mariacarla Gadebusch Bondio, Francesco Spöring, and John-Stewart Gordon

"The future was much better before."

—Karl Valentin, 1948

Knowledge of the future is a double-edged sword. Advances in pharmacogenetics and molecular genetics—together with societal expectations—have encouraged the development of numerous predictive procedures for the detection of predisposition to diseases. However, the knowledge that this affords physicians and patients is not necessarily only benign. The complex issues of medical prediction raise calls for a wide-ranging debate on the value and scope of foreknowledge (Habermas 2003). While medical *prognosis* has always been concerned with foretelling the probable course of a disease, *predictive medicine* is largely characterized by its focus on as yet unmanifested diseases or, in other words, on still "slumbering" pathological processes. It is based on temporal anticipation, on detection and prevention, on delay and control of such "virtual diseases," and so-called biomarkers play a crucial role. Taken as "objectively measurable characteristics" (Biomarkers Definition Working Group 2001), biomarkers have brought about a sea change in medical thinking, the results of which can already be seen. With the emerging paradigm shift in the ideas on illness and health through molecular genetics, these new entities carry within them information of a diagnostic, therapeutic, or prognostic nature.

In consequence, the quantity of sensitive biodata is continually growing. Biodata contain information about life, or more exactly, about the complex interdependency of life processes. To interpret this information, one should consider the whole network of affected organic connections within an individual being (Childs 2002). Most of these connections, however, are unknown today. While such connections will become the object of future research, the vision of a genetically based, new approach to health is widespread in industrialized countries. Although this popular perception has often been criticized as "genetic determinism" (Lewontin 1991; Bhardwaj 2006), several private companies offer "direct to consumer tests," or even the creation of individual health profiles based on genetic, genomic, and metabolomic examinations.

The consequences of the increasing focus on biomarkers are hard to foretell. Medicalization of life and increasing expectation regarding individual health literacy are merely two potential developments of a bundle of ethical issues. This bundle includes questions about

- the increasing responsibility of the "virtual sick," and on how public health systems ought to react to both probabilistic and deterministic predictions
- possible forms of discrimination and the warranty of data privacy
- potential psychosocial implications and the increasing tendency toward permanent health monitoring
- newly created necessities to make decisions, and on possible modes of communicating predictions
- the right to know, or, conversely, the right not to know

These and related issues—with the exception of potential discrimination and data privacy—are discussed in this book.

Recent Developments and Upcoming Challenges

Looking back over the last fifty years, it is clear that the advances in genetics and microbiology have led to an expansion of the foreseeable. Firstly, current predictions contain information on the probable occurrence of a disease as well as on predispositions (predictive biomarkers); secondly, they contain information on the probable course of an established disease (prognostic biomarkers); and thirdly, they contain information on the therapeutic effectiveness or success of a specific treatment (medicinal product-related biomarkers). In the last two decades, such predictions have had a significant impact on disciplines such as oncology (experimental therapy), pharmacology (in particular pharmacogenomics), psychiatry, epidemiology (see population screenings), human genetics, molecular biology, immunology, and so on. What these disciplines have in common is the search for biomarkers as interpretable data, which make it possible to determine diagnoses, predispositions, as well as the efficacy and potential side effects of drugs. Another commonality between these disciplines is that the predictions are usually framed by labels such as "individualized medicine," "personalized medicine," or "precision medicine" (Langanke et al. 2015; Vollmann et al. 2015).

In the so-called post-genome era, the potential of predictive areas of medicine has also been reflected in health policies. High hopes have been placed on the development of personalized prevention systems. Accordingly, research areas of medicine are particularly promoted if they study the period of time "before" patients show manifestations of disease (European Science Foundation 2012). The increasing medicalization of human life may result in health policy consequences that will be tangential to ethical debates on autonomy, non-maleficence, beneficence, and justice. Thus, it is of utmost importance to make the resulting consequences in terms of health policy the subject of an intensive, trans-disciplinary discussion (Habermas 2003; Gadebusch Bondio and Michl 2010).

So far, the philosophical and ethical implications of predictive knowledge on the one hand, and the clinical and individual implications on the other, have been separated from each other, and have only been studied fragmentarily. There is a lack of exhaustive interdisciplinary investigation, which does injustice to the complexity of the theme. This anthology aims to address this gap. The contributions collected here highlight the complexity and the implications of forward-looking medical knowledge. A central and guiding question is to what extent a change in mentality is necessary for the successful implementation of the predictive approach. At its center lies patient responsibility, which aside from patients also concerns physicians, researchers, families, and societies. The vast majority of literature on predictive and individualized medicine considers this change of mentality as a given. Most publications on personalized medicine assume that "people . . . are increasingly interested in being fully and accurately informed about the inherent changes in their health. An individual's life expectancy, quality of life, and life planning depend strongly upon this" (Golubnitschaja 2011). What must be questioned, however, is whether the preconditions for health responsibility exist in our society, and to what extent they can be optimized without creating new threats such as erosion of solidarity, societal coercion, as well as institutional, interactive, and indirect discrimination. While predictive knowledge can provide some orientation when facing health related uncertainties, its consequences for one's life planning nevertheless rely on existential considerations that determine quality of life, the value we put on it, as well as how we ascribe meaning to it. These subjective evaluations may never be fully grasped by means of quantification. It is in this sense that Guenda Bernegger et al. (2012) make the case for complementing probabilistic information with hermeneutic prognosis.

Increasingly, it will be the task of medicine in the future to provide individually tailored health profiles, which expands our "care for the self" (Foucault 1986). At present, the prognostic dialogue, in which both physician and patient interpret the probabilities of predictive information, is challenging (Rehmann-Sutter 2002). The transfer of generalized research results (e.g. epidemiological incidence of an illness) to individual cases (patient-related prediction) as well as the interpretation and communication of data and information provide broad scope for misinterpretations and misunderstandings (Gigerenzer 2014). However, strategies for reducing these have yet to be fully developed.

If prognosis stands as a paradigmatic example of uncertain knowledge in need of interpretation (with all its attendant hopes, fears, and doubts), its perceived value for the individual patient nevertheless remains very high. If a sick person puts herself in the hands of a physician, she wants to be informed of her condition in order to be able to assess whether and for how long she will be in position to arrange her own life. Patients must accept the risk of an imprecise prognosis and of failure. Tolerance limits vary in relation to the consequences that may follow fragile medical knowledge. Furthermore, these limits also depend upon factors such as culture and historical epoch.

Outline of the Book

The book is structured in three parts. All of them deal with the challenges that come with the rise of personalized medicine. However, they vary in their focus on different, yet overlapping spheres: namely, the sphere of individual challenges, of social challenges, and of research challenges. In Chapter 1, Mariacarla Gadebusch Bondio traces the historical development of medical thinking and argues that health has become a form of risk capital, while medicine's sphere of action has shifted toward anticipation and prevention, and carriers of genetic risks have increased responsibility for their own health. Increased patient responsibility is closely connected to mounting challenges in risk communication between patient and physician. With reference to cancer screening and cancer prevention, Ferretti, Linkeviciute, and Boniolo discuss the pitfalls associated with risk communication. They propose dedicated counseling, which could help patients to understand better the benefits and risks of each possible therapy.

John-Stewart Gordon elaborates on the "right not to know," and argues that ethical decisions—particularly ones in matters of procreation—cannot be justified on the basis of a presumed right not to know. Fangerau, Braune, and Lenk highlight crucial issues related to testing minors for genetic predispositions for late-onset diseases, and compare how existing guidelines for genetic testing deal with these concerns. Referring back to John Rawls' *A Theory of Justice*, they recommend not testing children for late-onset diseases—in particular if no therapy is available. Closely related to the "right not to know" are the issues arising in the context of incidental findings in genetic testing. After tracing the developing forms of genetic testing, Verena Steinke-Lange and Elke Holinski-Feder discuss these problems and propose several strategies for dealing with them.

The individualization of health responsibilities accentuates several challenges that need to be dealt with on a social level, for example, on the level of the health care system. Konrad Ott presents a risk-based argument as to why public health care systems should be based on the principle of solidarity. His argument relies on recognition of fellowship, on mutual trust, and on prudence. While Ott discusses a potential strategy for potential social conflicts, Peter Conrad and Miranda Waggoner describe an emerging development.

They introduce the concept of "anticipatory medicalization" using the example of "preconception care," with which medical surveillance is extended to virtually all women of childbearing age. Relating anticipatory medicalization to predictive medicine, they note potential consequences on the societal level.

While most chapters are based on "western" experiences, other contexts need to be recognized. Drawing on research in the fields of AIDS, malaria, diabetes, and cervical cancer, Steffen Flessa shows how prediction models may contribute to a transparent decision-making process in resource-poor countries. Xian-Ning Zhang and Ji Zuo make a case for acknowledging clinical genetics as a medical field in China. After illustrating the developments in Chinese policy regarding genomic technologies, they argue in favor of a liberalization of preventative eugenics. This is, of course, a controversial stance, but we include it as exemplary of how differently ethical challenges are perceived in different cultural contexts.

The third part of the book highlights specific potential for and difficulties of research in predictive medicine. Francesco Spöring describes the development of antidepressant drugs, and how the development of chemical agents, of diagnostic approaches, and of efficacy measurements has led to a narrowing focus on biomarkers. While acknowledging the potential for more refined patient stratification, he draws attention to the limits of quantification. In contrast, Stefan Lichtenthaler highlights the challenges of Alzheimer's disease (AD). After explaining the molecular causes of AD, he discusses potential treatment and prevention approaches currently in development, as well as the ethical issues that arise in the context of AD prediction. Simone Lista, Francesco Garaci, Nicola Toschi, and Harald Hampel focus on the role of biomarkers in early AD detection and prediction by discussing in detail the potential of qualified cerebrospinal fluid, magnetic resonance imaging, and multimodal neuroimaging. After this focus on neurological diseases, Agnieszka Pastula describes newer trends in cancer prediction. She discusses the prognostic and predictive value of immunoscore, circulating tumor cell counts, and of human-derived organoids.

At the interface between expectations regarding medicine and its epistemological and ethical limits, the book investigates how uncertain knowledge in medicine is presently dealt with. The anthology is deliberately and decisively interdisciplinary in design in order to create a foundation for transdisciplinary thinking (Mittelstraß 2001; Jahn 2008). This dialogue between clinicians and researchers, between the humanities and the natural sciences provides the basis for a creative examination of the tension between medical claims and the expectations of society.

Acknowledgments

The authors would like to thank Karl Hughes and Lucia Mair for their helpful comments and suggestions.

References

Bernegger, G., Musalek, M., and C. Rehmann-Sutter. 2012. "An Alternative View on the Task of Prognosis." *Critical Reviews in Oncology/Hematology* 84(S2):14–24.

Bhardwaj, M. 2006. "Looking Back, Looking Beyond: Revisiting the Ethics of Genome Generation." *Journal of Biosciences* 31(1):167–76.

Biomarkers Definition Working Group. 2001. "Biomarkers and Surrogate Endpoints: Preferred Definitions and Conceptual Framework." *Clinical Pharmacology and Therapeutics* 69:89–95.

Canguilhem, G. 1989. "Der Epistemologische Status der Medizin." In *Grenzen medizinischer Rationalität: Historisch-epistemologische Untersuchungen*, edited by G. Hermann, 69–93. Tübingen: Edition Diskord.

Chen, R., Mias, G. I, Li-Pook-Than, J., Jiang, L., Lam, H. Y., Chen, R., Miriami, E., Karczewski, K. J., Hariharan, M., Dewey, F. E., Cheng Y., Clark, M. J., Im, H., Habegger, L., Balasubramanian, S., O'Huallachain M, Dudley JT, Hillenmeyer S, Haraksingh R, Sharon D, Euskirchen G, Lacroute P, Bettinger K, Boyle AP, Kasowski M, Grubert F, Seki S, Garcia M, Whirl-Carrillo M, Gallardo M, Blasco MA, Greenberg PL, Snyder P, Klein TE, Altman RB, Butte AJ, Ashley EA, Gerstein M, Nadeau KC, Tang H, Snyder, M. 2012. "Personal Omics Profiling Reveals Dynamic Molecular and Medical Phenotypes." *Cell* 148:1293–1307.

Childs, B. 2002. "Medical Genetics to Genomic Medicine." *Medicina nei secoli* 14(3):707–19.

European Science Foundation. 2012. "Personalised Medicine for the European Citizen: Towards More Precise Medicine for the Diagnosis, Treatment and Prevention of Disease (iPM)." Available at http://www.esf.org/uploads/media/Personalised_Medicine.pdf [accessed January 27, 2016].

Foucault, M. 1986. *The Care of the Self*. Paris: Éditions Gallimard.

Fox Keller, E. 2000. *The Century of the Gene*. Cambridge: Harvard University Press.

Gadebusch Bondio, M., and S. Michl. 2010. "Individueller, Persönlicher, Präziser: Die Neue Medizin und Ihre Versprechen." *Deutsches Ärzteblatt* 107(21):1062–4.

Gigerenzer, G. 2014. "Breast Cancer Screening Pamphlets Mislead Women." *BMJ* 348:g2636.

Golubnitschaja, O. 2011. "Prädiktive, Präventive und Personalisierte Medizin—Eine Einführung." In *Personalisierte Medizin*, edited by W. Niederlag et al., 37–45. Dresden: Health Academy.

Habermas, J. 2003. *The Future of Human Nature*. Cambridge: Polity.

Jahn, T. 2008. "Transdiziplinarität in der Forschungspraxis." In *Transdisziplinäre Forschung: Integrative Forschungsprozesse verstehen und bewerten*, edited by M. Bergmann and E. Schramm, 21–37. Frankfurt/New York: Campus.

Langanke, M., Lieb, W., Erdmann, P., Dörr, M., Fischer, T., Kroemer, H.K., Flessa, S., and H. Assel. 2015. "The Meaning of 'Individualized Medicine': A Terminological Adjustment of a Perplexing Term." In *Individualized Medicine: Ethical, Historical and Economical Perspectives*, edited by T. Fischer, M. Langanke, P. Marschall, and S. Michl, 11–28. Cham/Heidelberg/New York/Dordrecht/London: Springer.

Lewontin, R.C. 1991. *Biology as Ideology: The Doctrine of DNA*. New York: Harper-Perennial.

Mittelstraß, J. 2001. "On Transdisciplinarity." *Trames* 15(4):329–38.

Pfleiderer, G., Battegay, M., and K. Lindpaintner, eds. 2012. *Knowing One's Medical Fate in Advance: Challenges for Diagnosis and Treatment, Philosophy, Ethics and Religion*. Basel: Karger.

Rehmann-Sutter, C. 2002. "Prädiktive Vernunft: Das Orakel und die Prädiktive Medizin als Erfahrungsbereiche für Rationalität." In *Zugänge zur Rationalität der Zukunft*, edited by N. C. Karafyllis and J. C. Schmidt, 202–32. Stuttgart/Weimar: J. B. Metzler Verlag.

Vollmann, J., Schildmann, J., Wäscher, S., and V. Sandow, eds. 2015. *The Ethics of Personalised Medicine: Critical Perspectives*. Surrey/Burlington, VT: Ashgate.

Wiesemann, C. 1998. "The Significance of Prognosis for a Theory of Medical Practice." *Theoretical Medicine* 19:253–61.

Part I

Individual Challenges

1 Beyond the Causes of Disease

Prediction and the Need for a New Philosophy of Medicine[1]

Mariacarla Gadebusch Bondio

In the play *Their Days are Numbered*, written in 1952, Elias Canetti (1905–1994) depicts a dystopian society in which instead of names, people have a number, the number of their years of life (Canetti 1999). Everyone knows the length of his or her life. Only their date of birth remains a personal secret.

"Ten," for example, is a child who is allowed to do everything. He does not go to school, he throws stones at people and tells lies. When a man called Fifty asks him why he does not want to learn or what he will do as an adult, he does not answer. The last question Fifty poses to him is decisive: "What's your name?" He answers: "My name is Ten," and Fifty understands why the boy behaves as he does (Canetti 1999, 202).

The play premiered in Oxford in 1956. It is difficult to say what induced Canetti to conduct this thought experiment. A world in which humans are predestined to live in the certainty of the moment of their death is a singular one. With the abolition of existential uncertainty, everything changes: instead of compassion or desperation when someone dies, resignation and enlightened acceptance are the usual reactions. The question that emerges when reading or watching Canetti's play, which he considered his most important work, is how life would feel if lived in the certainty of its unequivocal length.

It may be just a coincidence, but the play premiered at a time when a new approach was beginning to flourish in medicine. Between the 1950s and 1960s, concepts such as pharmacogenetics, the genetotrophic approach, propetology, and predictive medicine were introduced. These concepts were connected with discoveries in genetics, molecular biology, pharmacology, and biochemistry. After presenting some historical developments to help understand the growing need for a new philosophy of prediction in medicine, the challenges owed to the discovery of biomarkers will be discussed. Uncertainties and consequences connected with predictive information will be illustrated with reference to the genetic mutation BRCA1/2.

On the Modern History of Predictive Medicine

At the beginning of the 1950s, the structure of DNA had become the object of feverish research (Franklin and Gosling 1953; Wilkins et al. 1953). In

April 1953, James Dewey Watson (born 1928) and Francis Crick (1916–2004) published the news of their discovery of the molecular structure of DNA based on all its known features—the double helix (Watson and Crick 1953). Their model served to explain how DNA replicates and how hereditary information is encoded in it. Parallel to these seminal genetic discoveries, the biochemist Roger Williams published his concept of "genetotrophic disease." With this title, Williams and two colleagues announced their new approach to diseases in *The Lancet*.

> Based essentially upon recent findings in genetics and biochemistry which have not yet been incorporated into medical thought, the concept of genetotrophic disease may, we believe, lead to an understanding of many diseases whose aetiology is at present obscure.
>
> (Williams et al. 1950, 287)

What Williams and his colleagues presented here was a new concept that during the following years would be developed into a theory. Besides the genetotrophic concept, Williams developed the idea of so-called "biochemical individuality," to which he dedicated a whole book published in 1956 (Williams 1956).[2] Williams wanted to make clear that the traditional medical interest in norms and normal values in every single field of research was a distortion of reality:

> The existence in every human being of a vast array of attributes which are potentially measurable (whether by present methods or not), and probably often uncorrelated mathematically, makes quite tenable the hypothesis that practically every human being is a deviate in some respects.
>
> (Williams 1956, 3)

The ubiquity of variation in organisms and the significance of variability in humans had been facts, strictly speaking "a sine qua non of evolution" since Darwin; facts which could not be ignored if a serious understanding of human pathologies was to be sought (Williams 1956, 1). Although the importance of the individual constitution had already been noticed by the ancient physicians Hippocrates and Galen, this variability, according to Williams, had been neglected in biological sciences and medicine. He was convinced that the time had come to change medical thinking in this regard. Genetics was gaining ground quickly and conscious of this, Williams presented a systemic concept of disease capable of countervailing the impending threat of genetic determinism: According to an individual's genetic constitution, various aetiological factors with genetothropic origins had to be considered. These included cancer, diabetes, rheumatoid arthritis, disseminated sclerosis, and mental diseases as examples of pathologies in which the confluence of hereditary, "racial," and nutritional factors had been observed or could be supposed (Williams et al. 1950, 288). In 1961, he published a short, but no less programmatic article on "propetology," which he characterized as a "new branch of medical science" (Williams 1961, 325). What did Williams mean by this unusual term?

> The tools and the impetus to develop expertness in determining, in advance, what people are prone to contract any specified disease are now at hand; the new science needs only to be developed.
>
> (Williams 1961, 325)

Williams explains what had hindered the development of propetology until his time: an immature view of heredity and life was the reason for the rejection of the notion that many diseases have genetic roots. By means of the prevention of the development of mental impairment in babies prone to phenylketonuria, Williams illustrates how, by implementing diets low in phenylalanine, the effects of the pathology could be kept under control. The knowledge of the hereditary origins of this disease had helped to find a solution based on the management of environmental conditions. By propetology he was concerned to emphasize that people "merely lean toward" certain diseases without inevitably contracting them (Williams 1961, 325). Understanding the roots of resistance was a central aim: Why were some individuals able to escape the diseases they were prone to, while susceptible subjects became ill?

> People who are prone to have arthritis or stomach cancers or to become alcoholics or mental cases would not need to suffer from the respective diseases if the proneness were recognized early and if appropriate measures (largely unexplored up to now) could be taken to overcome it.
>
> (Williams 1961, 326)

The development of new measurements of "potential prognostic value" were the basis of this new field of medicine. Rather than standard tests, accurate and individualized prognostic research techniques were called for. At the end of his editorial, Williams announced a pioneer study in the area of propetology. The study was based on a series of tests on 500 young and healthy people over a period of years, and pursued the aim of identifying the types of maladies they were prone to. This study was characterized by a new methodological design: Individual patterns had been brought into focus and the concept of normality had been abandoned. In this regard, Williams' project was not singular by any means: The problem of norm and normality in medicine, in clinical practice, and more generally in biology was recognized by theoreticians and practical physicians (Murphy and Abbey 1967; Mainland 1969).

In Germany, this had been the topic of vehement debates since the 1920s on (Harrington 1996; Lawrence and Weisz 1998). Theodor Brugsch

(1878–1963) and other proponents of so-called constitutional medicine proclaimed the necessity of an individualized approach, of a "biology of the person" (Brugsch and Lewy 1926–1931; Kaiser and Hübner 1979; Brugsch 1986; Konert 1988). The reasons for the attention to individual variations were manifold. Between the First and the Second World War it had been observed that some people under similarly hard conditions (hunger, exhaustion, inadequate hygiene) fell ill and some did not. Of those who fell sick, some were able to recover, whereas others died. These were fundamental findings at a time in which therapeutic tools were limited. It was obvious to suppose that an inborn individual condition, such as a certain susceptibility to disease, had to play a role in the pathogenesis. From the 1920s on, studies in which the concepts of norms, normal values, and normality were questioned multiplied in German medical circles (Rautmann 1921; Grote 1922; Kaup 1926; Hau 2000; Timmermann 2001).

Discussions and critical views of normative tendencies in biology and medicine would find fertile ground from the 1950s onwards, both in the U.S. and in Germany (Ivy 1944; Murphy and Abbey 1967; Mainland 1969; Sunderman 1969; Gadebusch Bondio 2015).

One place where the discussions took place was in the emerging field of pharmacogenetics, at the intersection between genetics, biochemistry, and pharmacology (Humangenetik in Heidelberg 1991). Its pioneers were the Jewish-German-American physician and geneticist Arno G. Motulsky (born 1923), the German pharmacologist Werner Kalow (1917–2008), working in Toronto, and Friedrich Otto Vogel (1925–2006), a human geneticist and professor at the Freie Universität in West Berlin (Motulsky 1957; Vogel 1959; Kalow 1962). Motulsky, Kalow, and Vogel wanted to understand the role genetic factors played in the metabolism of drugs. They observed that only in cases of genetically susceptible individuals were specific drugs the specific agent that caused disease (drug reaction). Unexpected and unexplained reactions to drugs now could be explained. This made pharmacogenetics a discipline of importance for pharmacology, for therapeutics, and for medicine in general. But it took more than thirty years (until the late 1990s) to demonstrate the validity of the concept, and for it to be accepted by the scientific community. In his *Handbuch der Humangenetik*, Vogel dedicated one and a half pages to the "biochemical individuality of men" (with reference to Roger Williams), but considering the volume of the whole work (753 pages), the space allocated to it was negligible. While Williams emphasized the need to focus on individual variations, to remain reserved about the role of genes, and to maintain awareness of the complex array of extra-genetic factors interwoven with genetic factors, he maintained a critical and quite discriminating attitude to the emerging field of genetics (Vogel 1961, VI; Williams 1960, 96; Vogel and Motulsky 1979).[3] In an article dedicated to the history of pharmacogenetics published in 2002, Arno Motulsky put Williams' contribution into perspective: "He was admittedly not knowledgeable about genetics even though he strongly emphasized the role of heredity" (Motulsky 2002, 687). Motulsky's late critical judgment gives reason to suppose that Williams' genetotrophic approach had provoked skeptical reactions among the exponents of pharmacogenetics.

No list of the diverse pioneers of personalized and predictive medicine would be complete without Emanuel Cheraskin (1916–2001). Cheraskin was a physician and dentist. Following his monograph, *Predictive Medicine, A Study in Strategy*, which he wrote with W. Marshall Ringsdorf, the term "predictive medicine," which had already emerged in the 1960s, became programmatic (Cheraskin et al. 1966; Cheraskin and Ringsdorf 1973). In his immense publication output (approx. 700 titles!), Cheraskin insisted again and again on the multifactorial nature of health and disease (Cheraskin and Ringsdorf 1971). The classical, Hippocratic and Galenic idea of the gradual scale interposed between the limiting poles of health and disease is central to Cheraskin's concept of predictive medicine. Predictive medicine—according to Cheraskin—has its focus in the space between the endpoints of health and disease and "finds justification in the anticipation rather than simply in the identification of disease" (Cheraskin and Ringsdorf 1971, 511).

For Cheraskin and his colleagues, the research into predictive medicine becomes fundamental. The topic was discussed in the *Journal of the American Geriatric Society*, where Cheraskin and Ringsdorf both contributed with a series of articles on different aspects of predictive medicine in 1971. That medicine was in need of a new philosophy was now certainly recognized. The problem lay in identifying reliable indicators of health and disease which would enable prediction and allow preemption. Biochemical measurements had to be conducted within large groups of "presumably healthy" subjects of different "race," age, and sex (Cheraskin et al. 1966, 3). A higher incidence of cancer had been observed, for example, by subjects with a higher blood glucose level combined with advanced age and overweight measurement (Cheraskin et al. 1969). Cheraskin based his concept of predictive medicine on the predictive possibilities of the so-called "biochemical profile." He based his research program upon three theses:

> First, health and disease are a function of the interplay, in a product relationship, of host state and environmental stressors. Secondly, the oft-observed variability of response in terms of health and disease to similar environmental challenges is largely due to differences in constitution. Thirdly, host resistance and susceptibility is in part reflected and can to a degree be quantitated, by biochemical profile.
>
> (Cheraskin et al. 1966, 3)

The conviction that biochemical mensuration had predictive potential and prognostic value led the group, headed by Cheraskin, to combine traditional biochemical tests with new interpretations and understandings of physiological limits. A flavor of the still lively discussions about "normal" physiological values, and about what could be considered normal and healthy from a medical point of view, are recognizable when Cheraskin briefly describes the predictive value of his biochemical mensurations:

> The evidence suggests prognostic worth in utilizing traditional biochemical tests with conventional physiologic parameters; the data indicate greater predictive value employing traditional chemical procedures but with interpretation by presently unconventional physiologic limits. One could only speculate that more sophisticated physiologic standards derived from more sensitive biochemical instruments would likely further enhance predictability in the health sciences.
>
> (Cheraskin et al. 1966, 11)

In the following four decades, Cheraskin became one of the most popular nutritionists and a prolific author of best sellers (such as *Vitamin C: Who needs it?* or *Human Health and Homeostasis*). He never stopped trying to develop an operational predictive medicine program and stressing the need for a philosophy of prediction in medicine (Cheraskin and Ringsdorf 1973, 41–2). This was connected with the awareness of the need for new terminology. The redefinition of the concept of health and disease as a gradual and continuous alternation between the two poles maintained a central place in his medical theory.

During the 1950s, 60s, and 70s, these attempts to research pharmacogenetic and biomechanical variations laid important foundations for the individualized medicine of today. The players mentioned operated in different fields, but nurtured the vision of an anticipative medicine. The breakthrough came between the 1970s and the 1980s with the discovery of predictive biomarkers. This contributed immensely toward changing medical thinking about health and disease.

New Entities and New Ethical Challenges

Around 1975, a feverish search began for the correlations between quantifiable markers and the stage of cancer diseases, first and foremost breast cancer (Russel et al. 1975). One paper published by Joseph Paone and a group of oncologists in 1980 is regarded as the starting point for the introduction of the terminus 'biomarker', and for the beginning of the biomarker era (Paone et al. 1980; NIH Definitions Working Group 2000; Aronson 2005). The scientists discovered that a particular human serum was elevated in the presence of various neoplasms. In the clinical study, the serum levels were measured in randomized serum samples obtained from five different patient groups (from normal control persons to patients with benign breast disease, or with breast carcinoma in different phases). They found a positive correlation between the serum level and the clinical stage of breast cancer, coming to the following conclusion:

> In the present study, serum galactosyltransferase [GT] was found to be a sensitive marker for the presence of breast carcinoma. Furthermore,

> serum enzyme elevations correlated positively with the clinical stage of disease determined preoperatively. [. . .] The preliminary results of this study suggest that serum GT levels obtained sequentially, with time, may be potentially useful in the detection of early recurrence of breast carcinoma and as a marker for the measurement of tumor responses to therapy in patients with advanced disease.
>
> (Paone et al. 1980, 65)

The terms "marker," "biological marker," and "biomarker" became common currency. Today biomarkers have become the primary units in many branches of medical research. One of the current definitions of biomarkers is that of the U.S. Food and Drugs Association (FDA): "Biomarkers are a measurable diagnostic indicator that is used to assess the risk or presence of disease" (Biomarkers Definition Working Group 2001). In reality, the term refers to a broad subcategory of medically measurable and reproducible signs. This has produced a fundamental shift in the concept of disease. Biomarkers are considered "objective indications of medical state observed from outside the patient" (NIH Definition Working Group 2000). In contrast to medical symptoms, which are limited to those indications of disease perceived by the patient, a characteristic of a biomarker—following current definitions—is that it is "objectively measured and evaluated as an indicator of normal biological processes, pathogenic processes, or pharmacologic responses to a therapeutic intervention."[4] But—as Kyle Strimbu and Jorge Tavel (2010) noted—"The use of biomarkers in basic and clinical research as well as in clinical practice has become so commonplace that their presence as primary endpoints in clinical trials is now accepted almost without question. [. . .] In many cases, however, the 'validity' of biomarkers is assumed where, in fact, it should continue to be evaluated and reevaluated" (Strimbu and Tavel 2010, 463). Thanks to the "early data" provided by biomarkers, knowledge relating to open questions about risks and predisposition to disease, sensitivity to medication, and probable therapeutic reactions is increasing. This involves three radical changes in medicine and society: a) a new concept of health; b) a new understanding of medicine in terms of its aims; c) a growing responsibility for one's own health—in particular by carriers of genetic risks—and the dissemination of long-term surveillance procedures.

A New Concept of Health

Health becomes a fragile form of venture capital which is neither hidden in nor definable by the subject (Gadamer 1993). With the detection of risk-predictors in a state of being 'still healthy', the subject loses their power of definition as regards health. This 'expropriation' of health as a personal 'intimate' property has existential, social, and cultural consequences. The classical concept of the "cura sui"—with its long tradition from Seneca and Galen through to Foucault—implies the idea of the individual who looks after their health throughout their life (Foucault 2008, 1245). Over the centuries, medicine, moral-philosophy, and religion have all contributed to supporting the individual in shaping this physical and moral duty. Ancient physicians, such as Hippocrates and Galen, stressed the subjective character of health. Galen defined it as a "natural disposition, which causes actions," while disease was according to him an "unnatural disposition, which limits or prevents the same actions" (Galen, De san. tuenda, Buch 1, Kap. 5; Grimaudo 2006, 57; Gadebusch Bondio and Herrmann 2011). The transitions between illness and health were regarded by him as gradual and individually varied. This classical concept constituted the departure point for George Canguilhem's theory of "biological normativity" in the 1940s (Canguilhem 2011, 12). For him, being healthy meant being capable of defining and redefining one's norms depending on age and situation; it meant becoming ill and recovering. This constituted a sort of safety net for the here and now and a safety valve for the future. In short: a biological luxury. The current development in medicine challenges Canguilhem's concept of health dramatically. What is health in the era of prediction? A sort of threatened capital which can be neither valued nor disposed of. Medical surveillance and monitoring processes—both privately provided and state sponsored—is necessary in order to decide how to try to preserve it from risks.

A New Understanding of Medicine in Terms of its Aims

The new understanding of medicine is summed up by the program of the European Association for Predictive, Preventive and Personalized Medicine (EPMA), founded in 2008. According to the authors, what is involved here is a paradigm shift; medicine is moving from intervention to anticipation, in other words to prediction, prevention, and personalization. Predictive medicine has become the core of individualized medicine (Golubnitschaja 2015). Genetics and molecular biology have led to a steadily improving specification of what can be foreseen. The relationship between prediction and prognosis—in its classical sense—is also changing. The range of what is predictable has expanded to include: (1) the prediction of the probable occurrence of a disease (predictive biomarkers); (2) the prediction of the probable course of an existing disease (prognostic biomarkers); and (3) the prediction of the efficacy of a therapy (pharmaceutical biomarkers). Just how complexly intertwined these three levels of prediction are is shown by the Gene Mutation BRCA1/2. Depending on when this genetic variation is discovered—before the incidence of breast or ovarian cancer or in connection with the diagnosis of the cancer—different scenarios of possible intervention open up. This example leads to the third point mentioned above.

A Growing Responsibility for One's Own Health

The responsibility for one's own health, in particular for carriers of genetic mutations associated with increased risks for cancer, grows steadily. Without going into the complexity of a topic which involves oncologists, gynecologists, geneticists, and epidemiologists, I will try to define some emerging ethical problems. The strategies of prevention offered to the carriers of BRCA1/2 include the diminution of the probability of contracting the disease, the avoidance or the retardation of the incidence of the disease, and better and more effective therapies in order to increase life expectancy. Risk management has become a central tool of predictive medicine. Accordingly, one of the main duties of doctors is to be able to inform the carriers or patients about risks and probabilities, as well as about what can be done to prevent the worst outcome. Depending on which information is given and how, a certain medical choice can be made by the person concerned (Trepanier et al. 2004, 92). But how can the communication of probabilities enable individuals to estimate and understand their personal cancer risk, and promote the creation of a good decision-making process (Habbema and Hilden 1981; Hilden and Habbema 1987; Jacobi et al. 2009)? Over the past decades empirical and statistical models have been developed to estimate the risks of developing breast cancer. Risk assessment software has now been refined to include varying personal risk factors as well as family history and age. A consideration of all these aspects is necessary for reliable genetic cancer counseling. Only in this way can those giving advice do justice to their growing responsibility.

Taking Action against Risks

In 1990, the geneticist Mary-Claire King—a student of Arno Motulsky—reported in *Science* on the region of the genome in which she had detected the gene mutation BRCA1 (Hall et al. 1990). Between 1997 and 1998, the University of Utah and *Myriad Genetics* had developed the corresponding gene test, and then patented the gene segment BRCA1/2. They retained the monopoly on the gene test BRCA1/2 until June 2013.

Only a few weeks before (14th May 2013), Angelina Jolie's terse statement had appeared in *The New York Times* under the title "My Medical Choice." Gene mutation, and the attendant predisposition to developing breast and ovarian cancer, was suddenly brought to the attention of a worldwide audience (Paepke et al. 2009). When the Hollywood star encouraged women with commonly occurring familial cases of breast and ovarian cancers to have themselves tested and, if necessary, take preventative action, the gene test still cost between 3,000 and 4,000 dollars. Thanks to the landmark decision to forbid the patenting of the human genome handed down by the Supreme Court in Washington, genes may not be patented as "natural products." Only the artificially produced DNA copy, and the technique involved in producing it, can be patented. The movement which led to this sensational decision was a coalition of women's health organizations, patient groups, through to scientific associations, including Mary-Claire King, who took legal action against the U.S. Patent and Trademark Office and patent holders (Simoncelli and Park 2015). Meanwhile, Mary-Claire King has become active on behalf of population-based screening for BRCA1 and BRCA2, which should include all women regardless of their family history (King 2014).[5] This is in the firm belief that intensive monitoring together with previous findings can lead women to making informed decisions. The fact is that access to both the tests, and to the preventative offers in industrialized countries, is quite varied. Moreover, knowledge of the risks of a probable future illness, even with ideal health care provision, is uncertain knowledge. Dealing with 'future knowledge' on the one hand, and the consideration of preventative measures on the other, which can mean radical intervention in people who are 'still healthy', represents more than just medical challenges (Collins et al. 2014).

The medical literature alone is challenging. The studies which have been conducted on the examination of women with the BRCA mutations in the last two decades have in some cases produced contradictory results. The increase of follow-up and long-term studies is a result of beneficial developments over the last few years. The inclusion of different perspectives can also be observed: Alongside the patients, the physicians and nurses involved have also been the object of investigations. Difficulties in assessing risks, most of all in young women or women with moderate risk, remain a central challenge. In fact, one difficulty lies in developing evidence-based and accessible decision-support-tools capable of translating knowledge into clinical practice. For risk assessment of and therapeutic decision for a preventative mastectomy or ovarectomy with BRCA1/2 mutation carriers, and women whose family history is characterized by an increased risk of cancer, factors such as age and belonging to the various risk groups are fundamental. Dealing with predictive information and risk assessment is made more difficult for at least three reasons. **Firstly—different realities**: The results of international studies, including those in England, Australia, Israel, Canada, the U.S., France, and Poland, enable the observation of the role of different cultures, and above all of public health systems, as well as health care structures and financial conditions by the uptake of preventive measures (Metcalfe et al. 2008, 5).[6] In different countries and centers within countries, the uptake of risk-reducing surgery is highly variable. In 2009, the authors of a trial dedicated to the uptake of preventive mastectomy or ovarectomy by carriers of BRCA1/2 mutations, and by women whose family history was associated with increased cancer risks, came to the conclusion that long-term studies taking into account the age and the different risk-groups of the women concerned were still lacking (Evans et al. 2009). The remark that "careful risk counseling does seem to influence women's decisions for surgery, although the effect is not immediate" shows that psychological and emotional factors are crucial in decision-making processes concerning probable risks (Evans et al. 2009, 2318). One important conclusion was that the uptake of risk-reducing surgery is age-, risk-, and time-dependent, and that it is becoming higher overall. This is due also to the lack of alternatives. There is an urgent need for a broader range of non-surgical options. **Secondly—ethically necessary limitations in studies:** The long-term outcomes of risk-reducing surgery are positive overall. In a retrospective study including a group of 2482 BRCA1/2 carriers in twenty-two clinics and genetic centers in Europe and North America over thirty-four years (1974–2008), a drastic reduction of mortality and cancer incidence was observed in women who underwent ovarectomy and salpingo-oophorectomy (90% reduction of breast-cancer risk, 95% reduction of ovarian cancer, and 76% of mortality risks) (Domchek et al. 2010). But questions that remain open are the timing of risk-reducing interventions in relation to each other,[7] and in relation to the optimal age for the person undergoing risk-reducing surgery (Domchek 2010, 973). To evaluate risk-reducing surgery in relation to cancer risk and mortality reduction, a randomized controlled trial design would be necessary. But such a randomized approach in this field of research would not be ethically acceptable. Because of a lack of robust data, it is impossible to offer an evidence-based comparison of intensive breast cancer surveillance and risk-reducing surgery (Domchek et al. 2010, 974).[8] This difficulty is aggravated by contradictory results. A "survival analysis" published in 2010 in the *Journal of Clinical Oncology* comes to the conclusion that mammography and screening (MRT) are a reliable alternative to preventive surgery. The starting point for this research was the realization that the efficacy of prophylactic surgery and breast screening (annual mammography and magnetic resonance imaging) had never been empirically compared. The results show that although prophylactic mastectomy (PM) plus prophylactic oophorectomy (PO) at age 40 (yielding a 24% absolute survival gain for BRCA1 and 11% for BRCA2 mutation carriers) maximize survival probability, "substituting mammography plus MRI screening for PM seems to offer comparable survival" (Kurian et al. 2010, 222).[9] Due to these partially contradictory results, both uncertainty as well as the difficulty of personalized risk-management are increasing. **Thirdly—the role of the human factor in context:** Distortions in the interpretation of statistical data further complicate individual risk management (Singh et al. 2013). Even doctors have difficulty in understanding risk-related data: As Odette Wegwarth and Gerd Gigerenzer have demonstrated in a meta-study, data are often published in prestigious medical journals which are difficult to interpret and even more difficult to convey to the layman. Instead of absolute numbers (e.g. mortality rate ten in 1,000 cases), relative risks (mortality rate in 10% of cases) are reported, which are misleading, as it is unclear what the percent value relates to (Gigerenzer et al. 2010).

The cognitive obstacles are accompanied by the emotional stress for those women affected. A group of Californian female scientists has analyzed the factors that may influence decision-making regarding prophylactic surgeries among women with BRCA mutations (Singh et al. 2013). The study included 136 unaffected carriers of BRCA mutations, who either underwent prophylactic surgery or participated in a high-risk-surveillance program between 1998 through 2010. The examined subjects were divided into two cohorts, tested before and after 2005. This was the year when national guidelines were published regarding BRCA testing, and there were changes in insurance coverage for testing and prophylactic surgeries (Medicare) for women with familial disposition to breast cancer independently of age. These developments increased awareness of the availability of genetic testing. Among the carriers, 42% opted for a risk-reducing mastectomy, and 52% favored a risk-reducing salpingo-oophorectomy. The majority of the women who decided to undergo risk-reducing surgeries had a family history of breast or ovarian cancer; most of them had lost their mother to the disease, and were between 35 and 45 years old when surgery took place—on average two years after the genetic test. The authors came to the conclusion that besides the structural health care frame, the significance of psychological and emotional factors, and individual circumstances have to be considered when counseling at-risk women. Specialists have to understand that beyond the statistical risk, other dimensions are at least as important for directing personalized decision-making (Singh et al. 2013, e6)

Conclusion

The above-mentioned difficulties make Mary-Claire King's goal of having all women tested in relation to the BRCA1/2 gene variation appear to be a socially responsible measure in view of the increasing number of private providers of gene tests; however, it raises the question of which options potential gene carriers in various countries and contexts should be offered or not. If health in predictive, personalized, preventive, and now participative medicine is regarded as both capital worthy of protecting and a goal for preventive measures, then the various realities in the globalized world must be factored into the calculation to ensure broad, sensible, and effective application. For the industrialized countries in which predictive tests for the detection of gene mutation in women with family history are already established, the difficulties are shifting into the areas of predictive counseling and preventive decision-making.

Preventive decision-making is a dynamic process. Individual experience, family history, social status, and also the quality of medical advice and the access to valid information interact with each other. Comparative studies have shown that both the specificity of the care system with its particular preventive and prophylactic approaches, costs, facilities, etc. on the one hand, and the preferences of the treating physicians on the other, decisively direct the choice of the prevention pathway (Metcalfe et al. 2008).

One of the future tasks of medical ethics will be to create fully developed and practical models for predictive counseling. In other words, counseling concepts which enable a suitable response to the life history and life plan of those persons affected, and which take into account the fragile interplay between autonomy and vulnerability in virtually sick people. For molecular medicine, one of the central challenges is to remain receptive to the complexity of factors that influence pathological processes, and to realize that beyond the known causes of diseases, a myriad of cofactors interact. The vision of scientists like Roger Williams and Manuel Cheraskin outlined in the 1950s, and developed during the following two decades, was to grasp the nature of individuality. This meant going beyond the medical concept of the norm in order to understand 'headstrong' pathological processes, to foresee them, and to develop provision strategies to prevent them. The genomic revolution flanked by increasingly new technologies has opened up an array of predictive, diagnostic, and therapeutic tools. The old vision of a practicable predictive medicine is moving closer to its fulfillment. An historical review shows that today we are conjoined to talk less in terms of a paradigm shift in medicine, and more of the advance of a process which has been forgotten. For the precursors of individualized and predictive medicine, the enthusiasm bound up with the conviction that a new philosophy was necessary is justified and understandable from today's point of view. In view of the statistically difficult to determine probabilities of falling ill at a particular point in life, and the limited, indeed feeble prevention measures, the scope of intervention, and thus also responsibility, was limited. Thus, blithe talk of a paradigm shift is ill-advised. Awareness of past efforts to establish a form of medicine which enables an anticipatory rather than reactive approach, in other words, which is both predictive and preventive, puts such talk into perspective. Certainly we now have a far wider range of effective prevention measures than were available to the pioneers of predictive medicine. However, the increased choice that this affords brings with it its own dilemmas. In order to choose the most appropriate measures, statistical probabilities must be understood, assessed, and weighed up.

Managing uncertain knowledge concerning the future of genetically vulnerable persons demands a critical, and self-critical, attitude. In the context of predictive medicine, we can rarely say what is, only what could be if . . . different possibilities or strategies have to be assessed without knowing with any certainty whether and when something will happen. This fact forces physicians and concerned persons to imagine probable scenarios, to depict possibilities, and to communicate using conditional forms (if, whether). Medicine has increasingly to deal with probabilities, and to accept the limits of its fragile knowledge. It is not only bioethicists who suggest accepting this fundamental openness to future experiences as the basis for life-affirming and meaningful hope (Bernegger et al. 2012). A similar message was already to be found in Canetti's "Their Days are Numbered," written over half a century ago: Not knowing how or when one will die is preferable to knowing. It makes existence and the unforeseeable bearable.

Notes

1 I would like to thank Marion Kiechle, Myles Jackson, Stefan Lichtentaler, Francesco Spöring, Karl Hughes, and Ingo F. Herrmann for their valuable suggestions and their willingness to discuss the subject with me.

2 In 1998, forty years after its first edition, the book was to be rediscovered and republished. This testifies to medicine's growing interest in individuality and variations.

3 In Vogel's text book population, genetics also remains the central focus of research: Vogel, Friedrich: *Lehrbuch der allgemeinen Humangenetik*, Berlin, Göttingen, Heidelberg 1961, VI: "Zwei Gebiete sind es vor allem, auf denen in den letzten Jahren besondere Fortschritte erzielt wurden: die biochemische oder, wie man auch sagt, "molekulare" Genetik (die Analyse der Genwirkung) und die Populationsgenetik mit der Analyse der Mutation und der natürlichen Selektion. Hier musste deshalb der Schwerpunkt unserer Darstellung liegen. Gerade bei der Analyse der natürlichen Selektion etwa im Bereich der Hämoglobin-Varianten oder der ABO-Blutgruppenzeigt es sich, wie eng sich diese beiden Arbeitsgebiete miteinander verzahnen."

4 Biomarkers Definition Working Group (2001). Biomarkers and surrogate endpoints: preferred definitions and conceptual framework. *Clin. Pharmacol Therapeutics* 69, pp. 89–95; *Biomarkers In Risk Assessment: Validity And Validation*, published under the joint sponsorship of the United Nations Environment Programme, the International Labour Organization, and the World Health Organization, and produced within the framework of the Inter-Organization Programme for the Sound Management of Chemicals. World Health Organization Geneva, 2001, Available at http://www.inchem.org/documents/ehc/ehc/ehc222.htm [September 16, 2016]: "A biomarker is any substance, structure or process that can be measured in the body or its products and influence or predict the incidence of outcome or disease. Biomarkers can be classified into markers of exposure, effect, and susceptibility. If biomarkers are to contribute to environmental and occupational health risk assessments, they have to be relevant and valid. Relevance refers to the appropriateness of biomarkers to provide information on questions of interest and importance to public and environmental health authorities, and other decision-makers. The use of relevant biomarkers allows decision-makers to answer important public health questions by being used in research or risk assessments in a way that contributes useful information that cannot be obtained better by other approaches, such as questionnaires, environmental measurements, or record reviews".

5 In the case of women with a family history of cancer risk in Germany, not all health insurers accepted the costs of the test (under 1,500 Euro). The reason given is that the test is not a preventative examination, but merely determines predisposition to cancer.

6 Cf. Metcalfe (2008, 2017–22): "Assuming that cultural differences among patients within Canada are minimal, it suggests that cultural differences may not entirely explain the variations in the uptake rates—more likely were due to health care provider's recommendations and continuity of follow-up care. As expected, physicians have differing opinions on the effectiveness of various preventive options".

7 In a retrospective study including a group of 2482 BRCA1/2 carriers in 22 clinics and genetic centers in Europe and North America over thirty-four years (1974–2008), a drastic reduction of mortality and cancer-incidence was observed with women undergoing ovarectomy and salpingo-oophorectomy (90% reduction of breast-cancer risk, 95% reduction of ovarian cancer, and 76% of mortality risks). Cf. Domchek (2010).

8 "Intensive breast cancer surveillance does not reduce the risk of developing breast cancer, but aims to improve early detection. Due to our lack of detailed information on breast magnetic resonance imaging compliance, we cannot conclude that risk-reducing salpingo-oophorectomy improved breast cancer-specific mortality

compared with optimal screening; however, we found an association between women who underwent risk-reducing salpingo-oophorectomy and women who have better outcomes in terms of breast cancer risk, ovarian cancer risk, and ovarian cancer-specific survival, none of which would be anticipated to be affected by compliance to intensive breast cancer screening".

9 Kurian et al. (2010, 222): "Although PM [prophylactic mastectomy] at age 25 plus PO [prophylactic oophorectomy] at age 40 years maximizes survival probability, substituting mammography plus MRI screening for PM seems to offer comparable survival. These results may guide women with BRCA1/2 mutations in their choices between prophylactic surgery and breast screening".

References

Aronson, J.K. 2005. "Biomarkers and Surrogate Endpoints." *British Journal of Clinical Pharmacology* 59:491–4.

Bernegger, G., Musalek, M., and C. Rehmann-Sutter. 2012. "An Alternative View on the Task of Prognosis." *Critical Reviews in Oncology/Hematology* 84(2):14–24.

Biomarkers Definition Working Group. 2001. "Biomarkers and Surrogate Endpoints: Preferred Definitions and Conceptual Framework." *Clinical Pharmacology & Therapeutics* 69:89–95.

Brugsch, T. 1986. *Arzt seit fünf Jahrzehnten: Autobiographie*, 1st ed. 1957. Berlin: Verlag der Nation.

Brugsch, T., and F. Lewy, eds. 1926. *Die Biologie der Person: Ein Handbuch der allgemeinen und speziellen Konstitutionslehre in vier Bänden/Bd. 1 Allgemeiner Teil der Personallehre*. Berlin/Wien: Urban und Schwarzenberg.

Brugsch, T., and F. Lewy, eds. 1929. *Die Biologie der Person/Bd. 4. Soziologie der Person*. Berlin/Wien: Urban und Schwarzenberg.

Brugsch, T., and F. Lewy, eds. 1930. *Die Biologie der Person/Bd. 3 Organe u. Konstitution*. Berlin/Wien: Urban und Schwarzenberg.

Brugsch, T., and F. Lewy, eds. 1931. *Die Biologie der Person/Bd. 2 Allgemeine somatische und psychophysische Konstitution*. Berlin/Wien: Urban und Schwarzenberg.

Canetti, E. 1999. "Die Befristeten." In E. Canetti *Dramen*, 180–245. Frankfurt am Main: Fischer.

Canguilhem, G. 2011. *Le normal et le pathologique*, 1st ed. 1966. Paris: Presses Universitaires de France.

Cheraskin, E., and W.M. Ringsdorf. 1971. "Predictive Medicine: IV. The Gradation Concept." *Journal of the American Geriatric Society* 19(6):511–16.

Cheraskin, E., and W.M. Ringsdorf. 1973. *Predictive Medicine: A Study in Strategy*. Mountain View: Pacific Press.

Cheraskin, E., Ringsdorf, W.M., and D.W. Aspray. 1969. "Cancer Proneness Profile: A Study in Ponderal Index and Blood Glucose." *Geriatrics* 8(69):121–5.

Cheraskin, E., Ringsdorf, W.M., Setyaadmadja, A.T.S.A., and R.A. Barret. 1966. "Biochemical Profile in Predictive Medicine." In *Biomedical Sciences Instrumentation*, Vol. 3 (Proceedings of the Fourth National Biomedical Sciences Instrumentation Symposium, held May 16–18 in Anaheim, California, on Challenges to Medicine and Measurement), edited by J. Poyer, J. Herrik and T.B. Weber, 3–13. New York: Plenum Press.

Collins, I., Steel, E., Mann, G.B., Emery, J.D., Bickerstaffe, A., Trainer, A., Butow, P., Pirotta, M., Antoniou, A.C., Cuzick, J., Hopper, J., Phillips, K.-A., and L.A.

Keogh. 2014. "Assessing and Managing Breast Cancer Risk: Clinicians' Current Practice and Future Needs." *The Breast* 23(5):644–50.

Domcheck, S. M., et al." Domchek, S. M, Friebel, T. M, Singer, C. F, Evans, D. G, Lynch, H. T, Isaacs, C., Garber, J. E., Neuhausen, S. L., Matloff, E., Eeles, R., Pichert, G., Van t'veer, L., Tung, N., Weitzel, J. N., Couch, F. J., Rubinstein, W. S., Ganz, P. A., Daly, M. B., Olopade, O. I., Tomlinson, G., Schildkraut, J., Blum, J. L., Rebbeck, T. R. 2010. "Association of Risk-Reducing Surgery in BRCA1 or BRCA2 Mutation Carriers with Cancer Risk and Mortality." *The Journal of the American Medical Association* 304(9):967–75.

"European Association for Predictive, Preventive & Personalised Medicine." 2008. *The EPMA Journal*. Available at http://www.epmajournal.com [accessed June 15, 2013].

Evans, D. G., Lalloo, F., Ashcroft, L., Shenton, A., Clancy, T., Baildam, A. D., Brain, A., Hopwood, P., Howell, A. 2009. "Uptake of Risk-Reducing Surgery in Unaffected Women at High Risk of Breast and Ovarian Cancer Is Risk, Age, and Time Dependent." *Cancer Epidemiology, Biomarkers, and Prevention* 18(8):2318–24.

Foucault, M. 2008. "Der Gebrauch der Lüste." In *Sexualität und Wahrheit*, trans. U. Raulff and W. Seitter. In Id: Die Hauptwerke, 1154–370. Frankfurt a.M: Suhrkamp Verlag.

Franklin, R. E., and R. G. Gosling. 1953. "Molecular Configuration in Sodium Thymonucleate." *Nature* 171:740–1.

Gadamer, H. G. 1993. *Über die Verborgenheit der Gesundheit: Aufsätze und Vorträge*. Frankfurt am Main: Suhrkamp Verlag.

Gadebusch Bondio, M. 2015. "Das Individuum—eine Abweichung . . . und das Unbehagen der Wissenschaft." In *Norm als Pflicht, Zwang und Traum: Medizingeschichte im Kontext*, Vol. 19, edited by E. Brinkschulte and M. Gadebusch Bondio, :19–50. Freiburg: Peter Lang-Verlag.

Gadebusch Bondio, M., and I. F. Herrmann. 2011. "Ganz persönlich und doch so fremd—Gesundheit in Zeiten der Individualisierten Medizin." In *Was ist Gesundheit? Antworten aus Jahrhunderten*, edited by K. Bergdolt and I. F. Herrmann, 129–42. Stuttgart: Steiner.

Gigerenzer, G., Wegwarth, O., and M. Feufel. 2010. "Misleading Communication of Risk." *British Medical Journal* 341:791–2.

Golubnitschaja, O. 2015. "What This Book Series Is about." In *Individualized Medicine: Ethical, Historical and Economical Perspectives*, edited by Tobias Fischer, Martin Langanke, Paul Marschall, and Susanne Michl, v–vi. Cham/Heidelberg/New York/Dordrecht/London: Springer.

Grimaudo, S. 2006. *Difendere la salute: Igiene e disciplina del soggetto in "De sanitate tuenda" di Galeno*. Napoli: Bibliopolis.

Grote, L. R. 1922. "Über den Normbegriff im ärztlichen Denken." *Zeitschrift für die gesamte Anatomie* II(8): 361–77.

Habbema, J.D.F., and J. Hilden. 1981. "The Measurement of Performance in Probabilistic Diagnosis. IV. Utility Considerations in Therapeutics and Prognostics." *Methods of Information in Medicine* 20(2):80–96.

Hall, J. M., Lee, M. K., Newman, B., Morrow, J. E., Anderson, L. A., Huey, B., and M. C. King. 1990. "Linkage of Early-Onset Familial Breast Cancer to Chromosome 17q21." *Science* 250:1684–9.

Hampel, H., and D. Prvulovic. 2012. "Are Biomarkers Harmful to Recruitment and Retention in Alzheimer's Disease Clinical Trials? An International Perspective." *The Journal of Nutrition, Health and Aging* 16(4):346–8.

Harrington, A. 1996. *Reenchanted Science: Holism in German Culture from Wilhelm II to Hitler*. Princeton: Princeton University Press.

Hau, M. 2000. "The Holistic Gaze in German Medicine, 1890–1930." *Bulletin of the History of Medicine* 74:495–524.

Hilden, J., and J.D.F. Habbema. 1987. "Prognosis in Medicine: An Analysis of Its Meaning and Roles." *Theoretical Medicine and Bioethics* 8(3):349–65.

International Labour Organization and World Health Organization. 2001. *Biomarkers in Risk Assessment: Validity and Validation*. Published under the joint sponsorship of the United Nations Environment Programme, the International Labour Organization, and the World Health Organization, and produced within the framework of the Inter-Organization Programme for the Sound Management of Chemicals. Available at http://www.inchem.org/documents/ehc/ehc/ehc222.htm [accessed August 16, 2014].

Ivy, A.C. 1944. "What Is Normal or Normality?" *Quarterly Bulletin of Northwestern University Medical School* 18(1):22–32.

Jackson, M.W. 2015. *The Geneaology of a Gene: Patents, HIV/AIDS, and Race*. Cambridge: MIT Press.

Jacobi, C.E., de Bock, G.H., Siegerink, B., and C.J. van Asperen. 2009. "Differences and Similarities in Breast Cancer Risk Assessment Models in Clinical Practice: Which Model to Choose?" *Breast Cancer Research and Treatment* 115:381–90.

Jones, D.S. 2013. "How Personalized Medicine Became Genetic, and Racial: Werner Kalow and the Formations of Pharmacogenetics." *Journal of the History of Medicine and Allied Sciences* 68(1):1–48.

Kaiser, W., and H. Hübner, eds. 1979. *Theodor Brugsch (1878–1963): Hallesches Brugsch-Symposium 1978*. Halle/Wittenberg: Martin Luther Universität.

Kalow, W. 1962. *Pharmacogenetics: Heredity and the Response to Drugs*. Philadelphia/London: W.B. Saunders Company.

Kaup, I. 1926. "Bedeutung des Normbegriffs für die Personallehre." In *Die Biologie der Person: Ein Handbuch der allgemeinen und speziellen Konstitutionslehre in vier Bänden/Bd. 1 Allgemeiner Teil der Personallehre*, edited by T. Brugsch and F.H. Lewy, 191–225. Berlin/Wien: Urban und Schwarzenberg.

King, M.-C., Levy-Lahad, E., and A. Lahad. 2014. "Population-based Screening for BRCA1 and BRCA2: 2014 Lasker Award." *The Journal of the American Medical Association* 312(11):1091–2.

Konert, J. 1988. *Theodor Brugsch: Internist und Politiker*. Leipzig: S. Hirzel Verlag.

Kurian, A.W., Sigal, B.M., and S.K. Plevritis. 2010. "Survival Analysis of Cancer Risk Reduction Strategies for BRCA1/2 Mutation Carriers." *Journal of Clinical Oncology* 28(2):222–31.

Lawrence, C., and G. Weisz, eds. 1998. *Greater than the Parts: Holism in Biomedicine, 1920–1950*. Oxford: Oxford University Press.

Mainland, D. 1969. "Normal Values in Medicine." *Annals of the New York Academy of Science* 161(2):527–37.

Metcalfe, K.A., et al. 2008. "International Variation in Rates of Uptake of Preventive Options in BRCA1 and BRCA2 Mutation Carriers." *International Journal of Cancer* 122(9):2017–22.

Motulsky, A.G. 1957. "Drug Reactions, Enzymes, and Biochemical Genetics." *The Journal of the American Medical Association* 165(7):835–7.

Motulsky, A.G. 2002. "Pharmacogenetics: A Historical Account and Current Status." *Medicina nei secoli* 14(3):683–703.

Murphy, E.A., and H. Abbey. 1967. "The Normal Range—a Common Misuse." *Journal of Chronic Disease* 20:79–88.

NIH Definitions Working Group. 2000. "Biomarkers and Surrogate Endpoints in Clinical Research: Definitions and Conceptual Model." In *Biomarkers and Surrogate Endpoints*, edited by G.J. Downing, 1–9. Amsterdam: Elsevier.

Paepke, S., Schmid, R., Fleckner, S., Paepke, D., Niemeyer, M., Schmalfeldt, B., Jacobs, R.V., and M. Kiechle. 2009. "Subcutaneous Mastectomy with Conservation of the Nipple-Areola Skin: Broadening the Indications." *Annals of Surgery* 250(2): 288–92.

Paone, J.F., Waalkes, T.P., Baker, R.R., and J.H. Shaper. 1980. "Serum UDP-Galactosyl Transferase as a Potential Biomarker for Breast Carcinoma." *Journal of Surgical Oncology* 15:59–66.

Rautmann, H. 1921. "Untersuchungen über die Norm, ihre Bedeutung und Bestimmung." In *Veröffentlichungen aus der Kriegs- und Konstitutionspathologie 6:2, Heft* 2. Jena: Gustav Fischer Verlag.

Russel, H.D., Durie, B.G.M., and S.E. Salomon. 1975. "Polyamines as Predictors of Success and Failure in Cancer Chemotherapy." *The Lancet* 2:797–9.

Simoncelli, T., and S.S. Park. 2015. "Making the Case against Gene Patents." *Perspectives of Science* 23(1):106–45.

Singh, K., Laster, J., Karlan, B., Bresee, C., Geva, T., and O. Gordon. 2013. "Impact of Family History on Choosing Risk-Reducing Surgery among BRCA Mutation Carriers." *American Journal of Obstetrics and Gynecology* 208:329.e1–6.

Strimbu, K., and J.A. Tavel. 2010. "What Are Biomarkers?" *Current Opinion in HIV and AIDS* 5(6): 463–6.

Sunderman, W. 1969. "Computer Applications in Laboratory Medicine: The Delineation of Normal Values." *Annals of the New York Academy of Science* 161(2): 549–71.

Timmermann, C. 2001. "Constitutional Medicine, Neoromanticism, and the Politics of Antimechanism in Interwar Germany." *Bulletin of the History of Medicine* 75:717–39.

Trepanier, A., et al. 2004. "Genetic Cancer Risk Assessment and Counseling: Recommendations of the National Society of Genetic Counselors." *Journal of Genetic Counseling* 13(2):83–114.

Vogel, F. 1959. "Moderne Probleme der Humangenetik." *Ergebnisse der Inneren Medizin und Kinderheilkunde* 12:52–125.

Vogel, F. 1961. *Lehrbuch der allgemeinen Humangenetik*. Berlin/Göttingen/Heidelberg: Springer Verlag.

Vogel, F., ed. 1991. *Humangenetik in Heidelberg: Das Institut für Humangenetik und Anthropologie von 1962 bis 1990 im Lichte der Habilitationen* (Symposium am 10.03.1990 zum 65. Geburtstag von Professor Dr. h.c. Friedrich Vogel). Heidelberg: Universitätsbilbliothek.

Watson, J.D., and F.H.C. Crick. 1953. "Molecular Structure of Nucleid Acids." *Nature* 171:737–8.

Wilkins, M.H.F., Stokes, A.R., and H.R. Wilson. 1953. "Molecular Structure of Deoxypentose Nucleid Acids." *Nature* 171:738–40.

Williams, R.J. 1956. *Biochemical Individuality: The Basis for the Genetotrophic Concept*. New York: McGraw-Hill.
Williams, R.J. 1960. "Why Human Genetics?" *Journal of Heredity* 51:91–8.
Williams, R.J. 1961. "'Propetology', a New Branch of Medical Science? Editorial." *The American Journal of Medicine* 3(31):325–7.
Williams, R.J., Beerstecher, E., Jr., and L.J. Berry. 1950. "The Concept of Genetotrophic Disease." *The Lancet* 255(6599):287–9.

2 Comprehending and Communicating Statistics in Breast Cancer Screening. Ethical Implications and Potential Solutions

Giulia Ferretti, Alma Linkeviciute, and Giovanni Boniolo

Introduction

Cancer is among the leading causes of death around the globe, and breast cancer is the second most prevalent among women. Cancer screening and cancer prevention currently constitute the most promising way of tackling cancer in clinical practice and the public health domain.

The methods used for breast cancer screening are defined as *early detection*.[1] They are used to detect very early clues of the disease in order to treat it in its preclinical phase, and thus prevent symptoms and death. Note, moreover, that the majority of breast cancer screening programs for women who are not at increased risk due to genetic predisposition and family history commence at 50 years of age (Giordano et al. 2012), and it is still debated whether starting it earlier is a worthwhile undertaking.

It is generally believed that the strongest evidence supporting the claim that screening programs are beneficial is the reduced mortality resulting from the screened disease (Croswell et al. 2010). However, this claim tends to be taken for granted too often not only by physicians (Wegwarth et al. 2012) but also by policy makers (Croswell et al. 2010). Actually, evidence suggests that mammography, which is routinely used in screening for breast cancer, is not an effective option in reducing mortality in average risk patients (Stout et al. 2014). Moreover, about 90% of breast cancer cases are diagnosed in women who do not qualify as high risk patients and, therefore, should not be asked to participate in breast cancer screening programs (Stojadinovic et al. 2010).

Even though detecting breast cancer early offers the hope that starting treatment at its preclinical stage will result in a better clinical outcome, it may also lead to unwanted and potentially harmful outcomes such as *false-positive results* and *overdiagnosis* (Bretthauer and Kalager 2013). False-positive test results improperly indicate the presence of the conditions but are usually ruled out after additional diagnostic interventions. Nevertheless, even if finally proven mistaken, false-positive results engender unnecessary stress, anxiety, and health risks due to subsequent diagnostic exposure to radiation, biopsy, or even surgery. On the other hand, overdiagnosis refers to the detection of a condition that will not cause symptoms or death. Thus, overdiagnosed patients might be unnecessarily labelled with a lifelong disease, resulting in additional diagnostic tests, unnecessary treatment (overtreatment), and disease surveillance that is known to cause physical and psychosocial harm.

An alternative screening modality for breast cancer is magnetic resonance imaging (MRI), which is known to be more sensitive[2] and to detect cancers at an earlier stage. However, it also has significantly lower specificity as compared to mammography (Croswell et al. 2010), suggesting that its use would lead to more false-positive test results followed by unnecessary biopsies, and potentially unnecessary treatment. As a consequence, if breast cancer screening is carried out annually for a decade or longer, the cumulative risk of receiving at least one false-positive test result over a screening period might rise by up to 50% (Croswell et al. 2010). Such numbers are worrying, and might require major changes to how cancer screening programs are organized and promoted, which patients are offered the chance to participate, and how the information about potential benefits and risks is communicated to both individual patients and the public at large.

Our focus will be limited to individual patients, and we will advocate paying more attention to patient education and counselling. For example, concerning participation in breast cancer screening programs, patients should be informed about the possibility of false-positive test results and overdiagnosis, as well as further possible impacts on their physical and psychological health. This in turn could enhance patients' ability to make well-informed choices and reduce potential *decisional conflicts* (Biesecker et al. 2013).

By taking all of this into account, in what follows we will provide some advice on how patients could be helped in making well-informed choices about participation in breast cancer screening programs. We claim that specific counselling for patients could improve their decision-making process by explaining the concepts at issue, exploring the risks, benefits, and burdens of breast cancer screening, and emphasizing ethical considerations surrounding the screening practice.

Breast Cancer Screening: Tumor Biology, Statistics, and Uncertainty

It might be intuitively thought that detecting the disease early followed by immediate treatment might be a good strategy for improving survival and reducing mortality. This, in consequence, may result in claiming that screening programs are beneficial (Plutynski 2012). However, this could work only if early treatments are more effective as compared to treatments in later stages of the disease. Furthermore, reduction in mortality rates, even if age-adjusted, cannot be solely attributed to screening, but also to improved diagnostics, treatment, and surveillance, as well as to the increase in patients' awareness about the disease (Bretthauer and Kalager 2013).

In addition, numerous studies have shown that tumor biology can be very heterogeneous. This implies that carcinogenesis would be better described and explained through a *non-linear progression model*[3] than a *linear progression model*, which, therefore, renders consideration of future neoplastic steps more difficult (Croswell et al. 2010).

As previously mentioned, breast cancer screening methods are thought of as *early detection methods*: This means that very early and/or dormant forms of the disease are identified. However, the transition from the dormant to active stage of the disease has not yet been completely understood, and this could lead to biases when both the research data on screening tools and clinical information are interpreted. In particular, potential biases could arise when evaluating screening effectiveness (this includes lead-time/survival bias, healthy volunteer bias, length/disease development bias, and overdiagnosis bias).[4] Moreover, there is, on the one hand, the problem of base-rate fallacy,[5] whenever risks associated with screening are communicated, and, on the other hand, the problem related to the misunderstanding concerning the trade-offs between extended life and quality of life.

As we have seen, screening tests are more likely to detect dormant cancers, which in turn increases the risk of *overdiagnosis* (Brodersen et al. 2014). In fact, overdiagnosis can occur in two situations: (1) the detected cancer is benign, has no potential to grow, and could even spontaneously regress; (2) it might be a very slow growing cancer, which would never cause any symptoms in the patient's lifetime (Croswell et al. 2010). Therefore, it has been suggested that women undergoing breast cancer screening programs should not be referred to as patients but rather as participants, reserving the term "patient" for people with clinical signs, or symptoms of the disease (Bretthauer and Kalager 2013).[6]

Although it is known that overdiagnosis occurs, its precise definition is still disputed. A broad interpretation suggests that overdiagnosis has to do with the detection of abnormalities that would not have become clinically manifest in the absence of screening, and would have never caused disease symptoms or death (Bretthauer and Kalager 2013). Usually a more or less clear distinction is made between false-positive test results and overdiagnosis (Welch et al. 2011), specifying that the latter occurs when benign or slow-growing cancers are detected and treated while the former could be considered diagnostic mistakes which can be rectified after additional tests.

In the case of breast cancer screening, the estimates of overdiagnosis, occurring after mammography tests, range from 5% to 50% (Welch and Passow 2014). One review suggests that high estimates of overdiagnosis are due to the lack of adjustment to cancer risk and/or lead-time bias, and once these methodological flaws were resolved, the estimates might shrink to 5–10% (Puliti et al. 2012). Nevertheless, a meta-analysis of multiple clinical trials shows that the benefit of undergoing breast cancer screening for women under 50 years old may be very slight, and the cost in terms of false-positive test results, unnecessary biopsies, overdiagnosis, and subsequent overtreatment is relatively high (Mandelblatt et al. 2009).

On the other hand, various estimates suggest that two to two and a half lives are saved for each case of overdiagnosis (Paci 2012). This means that for every overdiagnosed and unnecessarily treated woman, there are about two or three patients who will benefit from early detection and treatment. Therefore, the false-positive result and overdiagnosis should be presented to the patients as potential screening outcomes by providing comprehensive statistical information in order to decide whether to take the risk of being overdiagnosed in exchange for a chance of being in a group that benefits.

Actually, attempts to eradicate the above-mentioned biases can be dated back to 1968. At that time, the World Health Organization (WHO) introduced the principles of screening (Bretthauer and Kalager 2013), which were subsequently developed into an analytic framework still used today (Croswell et al. 2010). Overall, the framework emphasizes that a screening should be justified according to (1) the seriousness of the disease screened; (2) the availability of effective interventions in an early stage; (3) the adequate understanding of the natural history of the disease (its development from latent to active stage included); (4) the fact that latent or early disease stages should be recognizable, and that examination methods should be acceptable to the public. Unfortunately, it seems that these requirements are not always respected in cancer screening programs.

As remarked, the major problems occurring in cancer screening are related either to the faults in the screening test, where optimal sensitivity and specificity are lacking (false-positive and false-negative results), or uncertainty when evaluating the further growth potential of benign and/or slow-growing cancers (overdiagnosis). The most common, but arguably not the most serious, burden of any disease screening is occurrence of a false-positive result, which can cause (1) *negative psychological consequences* such as anxiety, depression, changes in overall perception of personal health status, sense of vulnerability, which in its turn could significantly reduce quality of life; (2) *negative clinical implications* related to unnecessary subsequent diagnostic tests to confirm or reject the diagnosis, which expose the patient to the potential complications of additional interventions, such as biopsy or surgery; (3) *negative economic burden* for the patient and the health care system linked to the costs of the supplementary tests (Croswell et al. 2010). Concerning overdiagnosis, in addition to the above-mentioned burdens of false-positive test results, patients might also be exposed to lengthy, burdensome, and unnecessary treatments with potentially harmful side-effects.

All of this brings us to the next challenge, which is finding the best way to communicate the risks and benefits associated with screening and potential screening outcomes to individuals and the public.

Communicating Risks and Benefits

There is a shared consensus that individuals who are due to undergo a screening test aimed at otherwise healthy populations should be provided with decisional support in order to make an *informed choice*. There is less agreement, however, on what constitutes an informed choice, and which pre-screening interventions are the most effective in facilitating the process of truly empowered choice by patients. It is also not very clear how informed choice can be reliably measured. Despite these obstacles, it is suggested that informed choice has two core elements: (1) sufficient and balanced knowledge about the screening procedure and (2) attitudes reflecting personal values, which can be defined as thoughts or feelings toward the choice to be made (Biesecker et al. 2013). The former implies that decisions about participation in screening programs cannot be limited to evidence-based medicine alone (Croswell et al. 2010), as its scope is far wider than mere knowledge about risks and benefits. Incorporating individual patient values in the decision-making process brings it to much more sophisticated levels where we can also talk about what we refer to as a patient's *Personal Philosophy*. By this we mean "that wide set of more or less deep, coherent and justified metaphysical, methodological, religious, political, esthetical, ethical, etc., beliefs, assumptions, principles, and values that an agent possesses and that characterises in a unique way how he/she approaches the world and life. In other words, *Personal Philosophy* could be considered as the 'conceptual and value-laden window' from which any individual starts reflecting in order to make judgments, to make choices, and to act" (Boniolo and Sanchini 2016). Therefore, taking into consideration a patient's personal views, diverse backgrounds, and varying expectations could potentially improve communicating the risks and benefits of screening as well as help better to accommodate the patient-centered decision-making process.

Communicating statistical data might be a challenging task, especially when benefits and risks of cancer screening can be presented in many different ways. It is known that a patient's perception of cancer risk and screening benefit can be influenced by how information is presented. For example, more patients would choose to undergo screenings if they were informed about the losses of not being screened compared to patients who are presented with the possible gains (Klein and Stefanek 2007). However, empirical data evaluating the effectiveness of pre-screening interventions to enable more informed decisions remains scarce and unconsolidated, leaving it unclear whether and how decision aids affect value clarification as related to attitudes and decisions to participate in screening programs (Biesecker et al. 2013). Cancer screening constitutes a complex clinical practice, where the balance between benefit and risk requires both patients and physicians to face the problem of uncertainty. Therefore, grounding consent in trust and reciprocity rather than in information implies shifting the focus from the mere transfer of information to a more interacting form of communication, taking into account those aspects that are relevant for the patient's choice, such as the clinical circumstances and the patient's *Personal Philosophy*. It has already been shown in a research biobank setting (Boniolo et al. 2012; Sanchini et al. 2016) that shifting from informed consent to trust-based consent could be a way to overcome some difficulties surrounding informed consent when providing the patient and/or research subject with full and comprehensible information. This would emphasize the fiduciary role of health care professionals when screening-related decisions are made, requiring more personalized recommendations rather than generic application of screening guidelines.

Ethical Considerations Surrounding Breast Cancer Screening

Disease screening is considered an established public health measure in efforts to reduce mortality associated with a particular disease, which in our case is breast cancer. It has been shown that identifying and preventing the disease early in asymptomatic individuals works well on a population level, but it remains highly debatable whether undergoing screening is always worth the risk an individual patient has to take. Ethical examination of methodological approaches used in cancer screening programs, screening recommendations, and current screening practices reveal two leading considerations concerning, respectively, the patient's autonomy and beneficence (Plutynski 2012). Respect for a patient's autonomy refers to self-regulated actions, which are free from external control and limitations caused by inadequate understanding of the procedure. In contrast, adhering to the principle of beneficence entails contribution to a patient's welfare (Beauchamp and Childress 2013). Both are considered leading and established principles in biomedical ethics around which any health care service should be built.

Therefore, the participants in cancer screening programs should firstly be given full and comprehensible information, which allows them to make well-informed and autonomous choices in the light of existing uncertainties, potential risks and benefits of the procedure while taking individual values and personal philosophy into particular consideration. Secondly, the potential benefits and risks of screening should be presented in a balanced way, simultaneously clarifying the utility of screening procedures. Utility here refers to the effectiveness of a screening procedure in reducing the number of deaths associated with breast cancer, as well as the sacrifices which individual patients have to make. As stated above, some patients will have to undergo unnecessary treatment or further tests in order to support the infrastructure of saving lives in general. This presupposes that, on balance, some patients will be harmed while others will benefit from participation in screening.

Hence, breast cancer screening programs which are still widely promoted reveal that fundamental principles, suggested by the WHO nearly half a century ago, appear to be misunderstood, given low priority, or even considered irrelevant (Croswell et al. 2010). It is difficult to speculate about the reasons why individual risks associated with breast cancer screening are not always taken seriously and communicated to patients before the screening. One of the possible justifications for screening could be the fact that such interventions have proven to be beneficial at the population level, resulting in reduced mortality overall.

Nevertheless, as we have already mentioned, some biases do still occur. Overdiagnosis serves as a primary example, showing that detecting more cancers in their early stages does not constitute evidence that screening saves lives (Wegwarth et al. 2012). Moreover, individually based decisions can also raise ethical issues such as determining the best ways of communicating risks and benefits to the participants in cancer screening programs, and accommodating them for decision-making under uncertainty (Plutynski 2012).

It can be argued that participants in breast cancer screening programs, but in general for any cancer screening program, should be given the opportunity to decide for themselves whether to accept the risk of unnecessary post-screening tests or unnecessary treatment in cases of false-positive results or overdiagnosis. Of course, the problem of their fiduciary relationship with their health care professionals also arises here, and, thus, this is connected with the possibility of following their recommendations (Boniolo et al. 2012; Sanchini et al. 2016).

Breast cancer screening guidelines might work very well and offer clear benefit on a population level, but might not always be to the direct benefit of an individual patient. Therefore, in some cases cancer screening would violate the principle of doing no harm (non-maleficence) to the specific patient. Therefore, personalized care focusing on the individual situation rather than generic application of a guideline might be more beneficial and in line with respect for the patient's autonomy. It would also require individually tailored recommendations, more detailed consultations, and possibly also counselling sessions.

Improving Patient Support for Informed Decisions

In order to enable the participants in cancer screening programs to make empowered decisions, some features of screening programs have to be specified. For example, the sensitivity and specificity of the screening program should be clearly stated, as well as the prevalence of the screened disease in the general population. Moreover, participants should be informed about the risks of false-positive results and overdiagnosis (Plutynski 2012). Indeed, good communication aimed at informed decision-making about breast cancer screening should also include information about alternative screening modalities, such as breast ultrasound, MRI, and digital tomosynthesis, which are known to yield better specificity (Lee and Peters 2013).[7] However, they might not be available in all hospitals offering mammography screening, and also might not be covered by health care insurance, which would impose screening related expenses on the patient herself.

As already mentioned, questions of risks and benefits need to be supplemented by a discussion of the varying values patients attach to different risks, which could be better understood by examining the core elements of their *Personal Philosophy*. We would suggest that counsellors working with patients who are due to undergo cancer screening should be familiar with statistics and probability, and, in particular, with the different ways of presenting them to the patients because communicating statistical data can also be subject to potential pitfalls. For example, base-rate fallacy should be communicated with caution in order not to nourish the false impression that positive test results are meaningless. By the same token, it has to be made clear that a screening test alone does not constitute a final cancer diagnosis. There might be additional symptoms, in some cases also family history, and other tests which have to be considered in order to arrive at a more comprehensive and reliable clinical diagnosis. It is true that sometimes all additional diagnostic interventions are unnecessary, but this cannot be known beforehand. Therefore, information should be communicated as clearly as possible before the screening commences.

Figure 2.1 illustrates how the probability of a false-positive test result in breast cancer screening can be presented to the patient as part of written information or a visual aid during the counselling session. It helps the patient to understand what the 9% probability of a false-positive test means (however, attention should also be drawn to the fact that probabilities can be presented as absolute or relative, see Chapter 1 of this book). For example, if breast cancer prevalence in the general population is 1%, this means that 10 women out of every 1,000 can be expected to have a disease. However, the specificity of the test being 90% means that only nine of these women will test positive, and one who is affected by cancer will have a false-negative result. Furthermore, the 9% chance of a false-positive result means that out of 990 who do not have cancer, 89 will have a positive result, meaning that even though they do not have cancer, further investigation might be needed to rule out the possibility of the disease.

Caution is advisable when presenting statistics and their fallacies to patients, since it could also produce unintended consequences such as distrust in medicine and its methods in general. Indeed, to the extent that learning some statistics can be beneficial, they can also cause harm to the patient

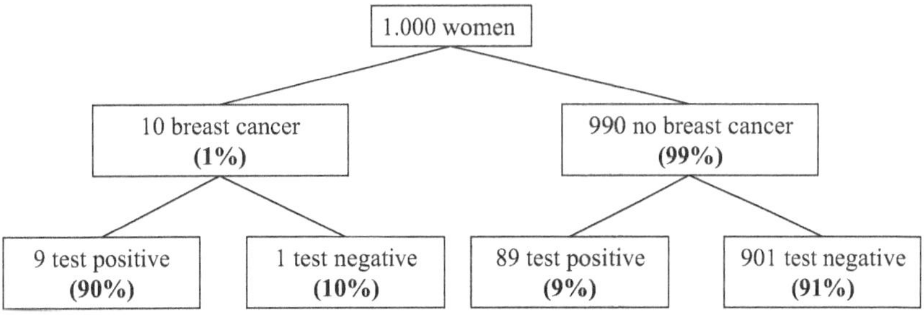

Figure 2.1 The probability of a false-positive test result in breast cancer screening explained in tree form, percentages, and frequencies. The graph also avoids *base-rate fallacy* while evaluating the benefit of screening.

if misinterpreted or used to feed wishful thinking or support denial of the diagnosis and failure to comply with follow-up appointments. This implies that skillful counselling is essential to enable patients to make well-informed and autonomous decisions, which are concordant with their personal values and beliefs, and at the same time serve their health interests best. Trust in health care professionals also plays a very important role, and can affect the decision, since patients' views on screening and overdiagnosis in general can vary. Some patients might still opt for screening following a "better safe than sorry" approach. Others, even if not fully understanding the information they receive, might have reduced intentions of undergoing mammography screening after being informed about the possibility of overdiagnosis (Waller et al. 2014), and they should not be left alone in the decision-making process.

As has been said, the data on the most effective communication and counselling methods remains inconsistent, but preliminary results are promising (Biesecker et al. 2013). Evaluation of decisional support intervention is complicated by the fact that in many cases there is no singular "best" choice which could be supported by concrete evidence that a certain screening, testing, or treatment option is better than the alternatives (Chiavari et al. 2015). In such cases, the best choice is dependent on the personal values and importance the patient attributes to the potential benefits, harms, and scientific uncertainties. Assistance in these situations can be offered through communication modalities such as decisional counselling (DeCo) and ethical counselling (EC). DeCo focuses on helping people to think about their choices by describing which choices exist, deliberating about potential options, and forecasting how they might feel about potential outcomes of the choices made (Chiavari et al. 2015). EC should be used to address the ethical issues involved in any decisional process, especially whenever probabilities and uncertainties surrounding the potential treatment outcomes are at stake. Moreover, exploring and reflecting on possible ways of action toward a specific intervention, such as breast cancer screening (Boniolo and Sanchini 2016), is also helpful. The developers of EC methodology see a patient as a privileged decision-maker where ethical issues are involved, recognizing the importance of communication, emotions, understanding of probabilities, and attributing high value to the patient's autonomy. EC is conceived as a non-directive, and serves as an orienteering tool when thinking about choices surrounded by ethical uncertainties. Its structure facilitates the guided reflection process where ethical arguments for supporting or rejecting each choice can be clarified and evaluated, taking personal values and beliefs into particular consideration, but not justifying one choice over another.

Conclusions

There are numerous considerations when breast cancer screening tests, such as mammography, are proposed. We suggest that dedicated patient counselling plays an important role in breast cancer screening care. Indeed, women should have the opportunity to make well-informed decisions in the light of comprehensive information about the screening test. Dedicated counselling services could be a viable solution to helping patients overcome and understand statistical uncertainties surrounding screening interventions. We have briefly described the main clinical and statistical puzzles creating uncertainties in breast cancer screening. We have also given an overview of possible harm which can result from overdiagnosis and false-positive results. Mammography screening saves two to two and a half lives in exchange for one case of overdiagnosis, which is positive on the population level, but can damage the individual patient. Therefore, the gap between the individual and the population levels needs to be filled by reformulating public risk communication, commissioning physicians to assess personalized risk based on a patient's clinical history, values, and socio-familiar role. We argue that individual values and preferences should be acknowledged when evaluating risks and benefits of cancer screening. Accommodating well-informed decision-making is the essential element in promoting a patient's autonomy, which is one of the foundational principles of clinical ethics. However, respect for patients' autonomy cannot be achieved without a proper understanding of the risks and benefits carried by the intervention. Therefore, clinical and epidemiological data should be communicated to the patients in a way which acknowledges their different personal philosophies and allows patients to attribute their own value to screening associated risks, benefits, and uncertainties.

Notes

1 There are two main methods used in cancer screening programs: (1) *prevention*, which detects and removes precursor lesion by preventing invasive disease and reducing its incidence; this means that cancer does not occur at all (this type of screening is used for colorectal and cervical cancers), and mortality from this disease is reduced; (2) *early detection* (used for breast, prostate, and lung cancer), whose aim is to detect cancer an early stage, and thus reduce site-specific cancer mortality by starting treatment early; it is beneficial if early treatment has been proven to be more effective.

2 Any disease screening test can be evaluated by two parameters: (1) *sensitivity*, which describes the probability that a negative test result means the absence of the disease, and (2) *specificity*, which describes the probability that the test will be positive if the disease is present.

3 Historically cancer was thought to advance along a series of progressive, increasingly dangerous steps. This presupposition led to the logical deduction that early detection, by breaking the chain of disease development, must be of benefit for an affected individual (Croswell et al. 2010). Later, however, it was shown that not all cancers progress in the same manner, and some might remain dormant in their latent stage for an entire lifetime.

4 These biases are relevant in the evaluation of screening programs because they can significantly falsify the results, and thus mislead policy makers in assessing health policy effectiveness and safety. The *lead-time/survival bias* refers to the added time of illness produced by the diagnosis of a condition during its pre-clinical phase. The *healthy volunteer effect* concerns screening in which the participants

are healthier than the general population spuriously showing increased benefits of the intervention. The *length/disease development bias* is typical of cancer screening because it refers to the inclusion in a survey of more slow-growing cancers with longer disease duration and better prognosis than fast-growing cancers.

5 The *base-rate fallacy* occurs when available statistical data on the general information are ignored in favor of specific data pertaining to certain information. For example, for every 1,000 women receiving mammography, only ten will have breast cancer (which is 1%), and nine of these will have a positive result (which is 90%). The remaining 990 women do not have breast cancer, but 89 will have a *false-positive result* (which is 9%). Out of the 98 positive results, only nine actually have breast cancer. The probability of having breast cancer given a positive test result equals to the number of positive tests of women with breast cancer divided by the total number of positive results (which is 9/89+9=9%).

6 We do not follow this division strictly. Therefore, the terms "patient" and "participant" in regard to cancer screening are used interchangeably.

7 This relatively new breast cancer screening method uses multiple x-ray images to create a three-dimensional picture of the examined breast. Currently available research results are promising, and it is thought that this screening modality might overcome the disadvantages posed by mammography screening.

References

Beauchamp, T.L., and J.F. Childress. 2013. *Principles of Biomedical Ethics*. Oxford: Oxford University Press.

Biesecker, B.B., Schwartz, M.D., and T.M. Marteau. 2013. "Enhancing Informed Choice to Undergo Health Screening: A Systematic Review." *American Journal of Health Behavior* 37(3):351–9.

Boniolo, G., Di Fiore, P.P., and S. Pece. 2012. "Trusted Consent and Research Biobanks: towards a 'New Alliance' between Researchers and Donors." *Bioethics* 26(2):93–100.

Boniolo, G., and V. Sanchini, eds. 2016. *Counselling and Medical Decision-Making in the Era of Personalised Medicine: A Practice-Oriented Guide*. Heidelberg: Springer.

Bretthauer, M., and M. Kalager. 2013. "Principles, Effectiveness and Caveats in Screening for Cancer." *The British Journal of Surgery* 100(1):55–65.

Brodersen, J., Schwartz, L.M., and S. Woloshin. 2014. "Overdiagnosis: How Cancer Screening Can Turn Indolent Pathology into Illness." *APMIS : Acta Pathologica, Microbiologica, et Immunologica Scandinavica* 122(8):683–9.

Chiavari, L., Gandini, S., Feroce, I., Guerrieri-Gonzaga, A., Russell-Edu, W., Bonanni, B., and F.A. Peccatori. 2015. "Difficult Choices for Young Patients with Cancer: The Supportive Role of Decisional Counseling." *Supportive Care in Cancer* 23:3555–62.

Croswell, J.M., Ransohoff, D.F., and B.S. Kramer. 2010. "Principles of Cancer Screening: Lessons from History and Study Design Issues." *Seminars in Oncology* 37(3):202–15.

Giordano, L., von Karsa, L., Tomatis, M., Majek, O., de Wolf, C., Lancucki, L., . . . Suonio, E. 2012. "Mammographic Screening Programmes in Europe: Organization, Coverage and Participation." *Journal of Medical Screening* 19(Suppl 1): 72–82.

Hersch, J., Jansen, J., Barratt, A., Irwig, L., Houssami, N., Howard, K., Dhillon, H., and K. McCaffery. 2013. "Women's Views on Overdiagnosis in Breast Cancer Screening: A Qualitative Study." *The BMJ* 346:f158.

Klein, W.M.P., and M.E. Stefanek. 2007. "Cancer Risk Elicitation and Communication: Lessons from the Psychology of Risk Perception." *CA: A Cancer Journal for Clinicians* 57(3):147–67.

Mandelblatt, J.S., Cronin, K.A., Bailey, S., Berry, D.A., and H.J. de Koning. 2009. "Effects of Mammography Screening Under Different Screening Schedules: Model Estimates of Potential Benefits and Harms." *Annals of Internal Medicine* 151(10):738–47.

Paci, E. 2012. "Summary of the Evidence of Breast Cancer Service Screening Outcomes in Europe and First Estimate of the Benefit and Harm Balance Sheet." *Journal of Medical Screening* 19(Suppl 1):5–13.

Plutynski, A. 2012. "Ethical Issues in Cancer Screening and Prevention." *The Journal of Medicine and Philosophy* 37(3):310–23.

Puliti, D., Duffy, S.W., Miccinesi, G., de Koning, H., Lynge, E., Zappa, M., and E. Paci. 2012. "Overdiagnosis in Mammographic Screening for Breast Cancer in Europe: A Literature Review." *Journal of Medical Screening* 19(Suppl 1):42–56.

Sanchini, V., Bonizzi, G., Monturano, M., Pece, S., Viale, G., Di Fiore, P.P., and G. Boniolo. 2016. "Research Biobanks: Why Information and Information-Based Consents Are Not Enough." *Bioethics*.30: 260–71. doi: 10.1111/bioe.12184.

Stojadinovic, A., Eberhardt, C., Henry, L., Eberhardt, J., Elster, E.A., Peoples, G.E., . . . C.D. Shriver. 2010. "Development of a Bayesian Classifier for Breast Cancer Risk Stratification: A Feasibility Study." *Open Access Journal of Plastic Surgery* 10:203–16.

Stout, N.K., Lee, S.J., Schechter, C.B., Kerlikowske, K., Alagoz, O., Berry, D., . . . J.S. Mandelblatt. 2014. "Benefits, Harms, and Costs for Breast Cancer Screening after US Implementation of Digital Mammography." *Journal of National Cancer Institute* 106(6):dju092.

Waller, J., Whitaker, K.L., Winstanley, K., Power, E., and J. Wardle. 2014. "A Survey Study of Women's Responses to Information about Overdiagnosis in Breast Cancer Screening in Britain." *British Journal of Cancer* 111(9):1831–5.

Wegwarth, O., Schwartz, L.M., Woloshin, S., Gaissmaier, W., and G. Gigerenzer. 2012. "Do Physicians Understand Cancer Screening Statistics? A National Survey of Primary Care Physicians in the United States." *Annals of Internal Medicine* 156(5):340–9.

Welch, H.G., and H.J. Passow. 2014. "Quantifying the Benefits and Harms of Screening Mammography." *JAMA Internal Medicine* 174(3):448–54.

Welch, H.G., Schwartz, L., and S. Woloshin. 2011. *Overdiagnosis: Making People Sick in the Pursuit of Health*. Boston: Beacon Press.

3 On the Nature of the Right Not to Know

John-Stewart Gordon

Introduction

In the last three decades, there has been a growing interest in and (most recently) a lively debate about the idea that there is "a right *not* to know." This right is allegedly based on the principle of autonomy, and its advocates consider it a parallel to the well-known, autonomy-based right to know, which is universally granted in the context of informed consent in medical ethics (e.g., Chadwick et al. 1997; Rehmann-Sutter 2009; Knoppers 2014). In this article, however, I do not intend to go through all the various positions[1] on this issue held by well-known authors (e.g., Ost 1984; Husted 1997; Andorno 2004; Malpas 2005). Rather, I will confine myself to one line of argumentation that has been mentioned in the debate, but not properly examined previously (see Macklin 1992; Nuffield Report 1993).[2] My general claim is that the supposed right not to know must be *absolute* by nature, if it is going to be able to carry all the weight that proponents need for their particular ethical reasoning and decision-making. Since it is questionable whether absolute rights exist at all, I argue that ethical decisions—particularly ones in matters of procreation (e.g., Malek and Daar 2012)—cannot be justified on the basis of a presumed right not to know. Rather, the very idea of an absolute[3] right not to know is self-refuting.

This result has important implications for the relationship between "the right not to know" and the new technological possibilities of predictive medicine. For example, if people do have a right not to know, then predictive medicine might become useless in cases where the patient refuses to be informed, even though the information might save his or her life, a partner's life (e.g., from death or serious illness due to HIV transmission), or the lives of possible future offspring, to whom a fatal genetic disease may be passed. In that respect, genetic tests make sense only if they are used for some purpose—one should not perform tests simply for the sake of testing. That is why, for example, there is a close relationship between genetic testing and the selection of healthy embryos that do not carry any genetic diseases, or are not at risk of severe impairment. The very idea of embryo selection is itself hotly contested by many scholars and activists involved in disability studies on the basis that it constitutes discrimination against people with impairments. These and other complex issues call for a careful examination of the nature of the right not to know.

Cases

The Soccer Case

Brian loves soccer, but due to some unforeseen circumstances, he was unable to watch his favourite team play in the semi-final round at an important soccer championship. He asks his roommate not to tell him the results of the game, because he wants to watch it on playback and enjoy a nice evening. Unfortunately, his friend intentionally tells him the results. Can this be considered a violation of the so-called right not to know?

HIV Case

Bob has been tested for HIV because he wants to donate blood. He is known to engage in unprotected sex and also strongly refuses to hear any bad news about his health condition. Unfortunately, the test is positive and the physician is wondering what he should do, given the patient's statement that he does not wish to be informed of any bad news concerning his health. Is the physician allowed, or even morally obligated, to tell Bob the test result, given that he knows that doing so would be against the patient's wishes? Could telling him the result be considered a violation of the right not to know? Furthermore, if the physician is not obligated to reveal whether Bob's donated blood will be used, should he still tell Bob the truth about his health condition?

The Chorea Huntington Case

Lastly, a married couple, Peter and Laura, want to have a child, but the couple knows that Peter has a family history of Chorea Huntington (CH)[4] and that, if he carries the gene, the chances are 50% that the couple's child will have the disease. Peter does not know whether he actually has CH and does not want to be tested for it. Do parents have an obligation to be tested for CH (or other potentially lethal or debilitating diseases) if they want to procreate and know that there is a family history of that disease (see Malek and Daar 2012[5])?

Many countries, for example, lack the technological capacity to provide parents with pre-implantation genetic diagnosis (PGD) or a non-disclosure PGD (in case the husband does not want to know his own condition) for the purpose of determining whether the fetus carries the disease or not. (I am not even talking about the issues of very high cost, exhausting test processes, and considerable psychological burdens; some parts of the world are not equipped to provide the testing and diagnosis at all.) For the sake of argument, however, let us assume that the only possibility in our case is to directly test the fetus's DNA for the CH mutation.[6] The parents refuse to have this test because, should the test be positive, that information would reveal that the at-risk parent does carry the disease, which is something that the husband does not want to know.[7] Suppose that they want to procreate without having any test because getting the test results could undermine Peter's supposed right not to know. Is there a moral duty to refrain from having children in such cases (e.g., Asscher and Koops 2010, 32–3, who believe that denying the couple the right to have children is an extreme position that cannot be seriously promoted; against this view see Macklin 1992; Malek and Daar 2012)? What is the morally right thing to do?

A Brief Response

Concerning the first case, the soccer game, it seems unreasonable to claim that the friend was in fact violating Brian's right not to know by revealing the results of the soccer game. We would say that Brian's friend, rather unkindly, ignored Brian's stated desire, but we would not invoke any conception of rights. Revealing the truth spoiled Brian's opportunity to enjoy watching the game on playback, but it certainly did not violate a moral or legal right.

The second case (involving the HIV test) and the third case (CH) are different because they involve the possibility of serious medical harm to other people—that is, to sex partners and offspring. It seems that, in both cases, two important moral principles or rights clash: the principle of nonmaleficence (i.e., do no harm) and the right not to know, which is (according to its proponents) based in the principle of autonomy. These two cases are obviously different from the first case because a third party might be seriously harmed by having unprotected sex with someone who has HIV, or through the process of procreation when one at-risk partner might have a lethal disease such as CH (see Macklin 1992; Nuffield Report 1993).

Objection—A Life with CH is Non-tragic

It might be objected, however, that CH or simply a predisposition for CH does not imply a life that is unliveable or even tragic in any way. Peter or his potential offspring might, theoretically speaking, become ill, but they could still live valuable and good lives up to that point without knowing their vulnerability to a specific disease. At first sight, this objection seems to carry some plausibility by arguing that all is good and fine if only we do not know about our future fate, but at second glance this claim proves to be seriously misleading and quite naive. The simple reason is that Peter (and Laura) will always live with the fear that Peter and/or their offspring might die an early and unpleasant death—a fear that usually brings with it a great psychological effect on the persons impacted. In addition, the objection overlooks the strong possibility that Peter would be unable to maintain responsibility for his family if he dies at an early age. Third, the very experience of witnessing Peter's early and unpleasant death is most certainly a great burden for his family, which should be taken into account as well. It is not argued that Peter (or persons in similar situations) should not be allowed to have a partner or family, but it is certainly reasonable to consider how his early death will be received by his beloved ones—especially if the offspring might share the same unpleasant and lethal fate in adult life, given the high chances of passing on to the next generation a fatal, incurable disease characterized by early physical and cognitive deterioration.

For example, according to the German Gendiagnostic Law (GenDG) of 2009/2010, § 15.2:

> Eine vorgeburtliche genetische Untersuchung, die darauf abzielt, genetische Eigenschaften des Embryos oder des Fötus für eine Erkrankung festzustellen, die nach dem allgemein anerkannten Stand der medizinischen Wissenschaft und Technik erst nach Vollendung des 18. Lebensjahres ausbricht, darf nicht vorgenommen werden.[8]

The idea underlying this regulation in German law is that parents should not make decisions based on how a disease may affect their offspring's own adult lives, regardless of whether the disease is fatal and/or incurable. If Peter and Laura agree to test Peter for CH before they procreate—the results of which might, for the sake of argument, influence their decision to have an offspring—the German law does not apply. Of course, on one hand, it is impossible for the state to intervene and tell the parents that they may not be tested for CH (as being tested is generally a good thing) simply because a positive test result might discourage them from procreating. On the other hand, however, it seems morally irrelevant whether Peter or an embryo (which is commonly not seen as a person with full moral and legal rights) is tested for CH. It seems idiosyncratic and highly dubious for German law to make it impossible for future parents not to base their decision to procreate on the genetic results of an embryo (or fetus), if the parents want to avoid having an offspring who will develop a fatal and incurable disease in early adulthood.

Of course, given the particular history of eugenics programs in Nazi Germany, it is understandable that the very idea of genetic testing, along with its possible consequences for embryos and fetuses, is a complex and sensitive topic in contemporary Germany (e.g., in the context of disability). Nonetheless, philosophically speaking, the German regulation cited here seems rather pragmatic and context-specific, and does not do justice to parents' autonomy concerning their family planning. This tendency becomes even more visible when we observe the contrast with cases of genetic predispositions of possible offspring that lead to a fatal and incurable disease in early childhood rather than adulthood. Here, German law permits parents to obtain the results of a test performed on the embryo or fetus, and then decide whether to carry the pregnancy to term, depending on the test results. It seems highly unreasonable to distinguish morally between the two types of cases according to when the fatal and incurable disease will occur. Where do we want to draw the line?

In what follows, I will examine directly the argument for grounding the right not to know in the principle of autonomy. However, before I deal with this important issue, it seems reasonable to provide some details about the historical development of the right to know in the context of medical ethics, because this is the background against which the lively debate concerning the so-called right not to know has emerged during the last three decades. As I have hinted earlier, however, my contention will be that the right not to know cannot be based on the principle of autonomy.

The Rise of the Principle of Autonomy in Medical Ethics

History

The idea of paternalism in medical ethics, probably best exemplified by the statement that "the physician knows best," was considered a truism from the times of the Hippocratic Oath in ancient Greece until relatively recently. The very idea that patients have a say in or should be informed about their medical treatment was completely absent until the beginning of the last century. The rise of autonomy in the context of the physician-patient relationship can thus be seen as the counter-movement to paternalism in health care. The concept of personhood and the principle of autonomy pervaded debates over medical ethics, significantly shaping discussions both in academia and on the ward (see, Gordon 2012a).

The initial demand for medical ethics was largely in reaction to some serious incidents of abuse, such as the research experiments on human subjects conducted by the Nazis and the Tuskegee Syphilis Study (1932–1972) in the U.S. At that time, bioethics was essentially driven by extreme cases, and did not consider systematic problems in health care, such as the issue of access to quality care. However, in reaction to these horrible events, the Nuremberg Code (1947) and the Declaration of Helsinki (1964) were created in order to provide researchers and physicians with ethical guidelines with regard to respecting the autonomy, or (as it is usually referred to in modern health care and research) the individual informed consent of patients and research subjects (see, Gordon 2012a).

However, we should acknowledge earlier advances in recognizing patients' right of autonomy. For example, "the [very] idea of individual informed consent is due to the Prussian and German bureaucratic regulations of 1900–1901 that appeal to the case of Dr. Albert Neisser, who in 1896 publicly announced his concern about the possible dangers to subjects whom he was vaccinating with an experimental immunizing serum" (Zentralblatt der gesamten Unterrichtsverwaltung in Preussen 1901, 188). Thirty years later, the investigation into the death of 75 German children caused by the use of experimental tuberculosis vaccines in 1931 revealed that mandatory informed consent had not been obtained (Rundschreiben des Reichsministers des Inneren, 28 February 1931, in Sass 1989, 362–6). Against this background, Baker rightly states that "the informed consent doctrine was thus originally a regulatory innovation created by Prussian bureaucrats; it was not an artefact of American legal or philosophical culture, but of German bureaucratic culture. It was a German solution to problems created by the advances of German biomedical science" (Baker 1998, 250; see Gordon 2012a).

The Notion of Autonomy

Since Beauchamp's and Childress's pioneering work *Principles of Biomedical Ethics* (2009), originally published in 1977, the idea of individual informed consent has become an influential and vital part of the principle of autonomy as applied to medical ethics. As we have seen, the notion of informed consent was known long before that time, but Beauchamp and Childress were among the first to use this concept systematically for bioethical reasoning and decision-making, giving it a prominent place in their theory.[9] According to them, a patient is capable of autonomous, well-informed decisions in medical contexts if their procedure determines that the patient is competent. If the particular patient does not pass the threshold and information elements of the procedure in question, then he or she is unable to give informed consent regarding a medical treatment, and hence must be considered incompetent.[10]

Beauchamp and Childress (2009, 120–1) claimed that there are three main components of informed consent. The first consists of the threshold elements used to determine whether a person generally meets the preconditions of autonomous decision-making. This component contains two parts, namely, being competent (i.e., able to understand and decide) and the voluntariness condition (i.e., having the freedom to decide without being coerced by an external party). The second main element concerns three parts that deal with issues of information: the disclosure of material information, the recommendation of a treatment plan, and the understanding of all relevant information in the particular case at stake. The last main element is called the consent element and comprises two parts: the final decision in favor of or against a particular treatment plan, and the authorization of a chosen treatment in case of a positive decision. Beauchamp's and Childress's idea of how to speak about informed consent influenced the fields of medicine and medical ethics all over the world and was the gold standard for several decades.

In brief, the idea of basing the right to know on the principle of autonomy is the result of the recent history of the rise of patients' rights and the decline of paternalism in the 20th century. To reveal to a patient the details of his or her health status, and to let the patient partake in all decisions related to his or her health has become obligatory. In other words, the patient has a moral and legal right to be informed about his or her health status and possible treatments. The principle of autonomy is the very foundation of this particular right. Beauchamp and Childress contended, however, that the principle of autonomy is a *prima facie* and not an *absolute* principle. That means that in some cases of conflict with other principles, the principle of autonomy might be overridden in favor of other important principles, such as the principle of non-maleficence.

In the next section, I will examine an example of such a substantive clash between principles and the consequences of upholding the idea of a so-called right not to know.

Basing the Right Not To Know on Autonomy

A Self-Refuting Endeavor?

Living with an incurable lethal disease is highly burdensome, at times severely painful, psychologically nerve-wracking, physically and mentally extremely demanding, and usually something one does not want for others, not to mention for one's own offspring. Nevertheless, some parents, such as Peter and Laura in the case cited above, may want to procreate despite uncertainty about whether they could pass on an incurable, lethal disease to their offspring. They may choose to avoid becoming informed about their health status by refusing to take a diagnostic test; this refusal is, in their view, justified by the right not to know. The right not to know, however, is eventually based on the principle of autonomy, which is commonly seen as the very foundation of human action.

What about the inner logic of this line of argument? If one claims that the right not to know is based on the *prima facie* principle of autonomy, and can therefore supersede any other possible principles (such as the principle of non-maleficence) that might come in conflict with the right not to know, then one must necessarily hold either that the principle of autonomy is an *absolute* principle that can justify the right not to know against *all* other possible conflicting claims, or that the *prima facie* principle of autonomy is justifying an *absolute* right. Neither argument is convincing, and I see no way to provide a solid justification for either of these extreme claims.

Proponents of the right not to know must hold the view that this right always prevails in cases of conflict. They cannot claim that the right not to know is a *prima facie* right, because then they would allow for the possibility that they might be unable to always justify their own decision in cases of conflict between principles (such as between autonomy and non-maleficence).

It is a matter of debate whether there are *any* absolute rights or principles. Regardless, there are currently no convincing and watertight arguments for identifying any specific principle as absolute. Furthermore, a proposal that the right not to know might deserve status as an absolute right is particularly unconvincing and, given the complexity of the moral life in the field of medical ethics, indeed self-refuting.

Objection—On the Reliability of First-Person Reports

There is an analogous and very common argument that has also been used in the context of disability and debates over quality of life that adheres to the reliability of first-person reports. It runs as follows: Since healthy people are unable to truly put themselves into the shoes of a person with a severe impairment, they assume without further argument that the quality of life of a person with a severe impairment is less good; depending on the severity of the impairment, they might even call such a life a *wrongful life*, i.e., one that is not worth living. One might take this position in the case of anencephalic newborns, or human beings with an extremely limited mental capacity. Additionally, many people argue that people with paraplegia live in some sense a less valuable life, even though many scholars and activists in disability studies, citing empirical data, reject this claim. In other words, many people with severe impairments place greater value on the quality of their own lives than non-impaired people attribute to them. As Ron Amundson puts it: "The question is this: Who should we believe? Should we assume that first-person reports are more likely correct than judgments made by others? If not, when is it appropriate to doubt first-person reports about quality of life?" (Amundson 2010, 374). Scholars such as Jerome Bickenbach have suggested—and I strongly agree with him—that people with severe impairments might simply be mistaken in their judgement of their own quality of life, given that the good life most probably consists of a combination of subjective and objective criteria.

Given these considerations, let us move on to the (analogous) argument that we are unable to determine the quality of life of a person who suffers from a fatal and incurable disease. Since (the objection begins) we approach the situation from a perspective external to the healthy person, it is impossible to be non-judgemental since ill health is *a priori* considered something bad in itself that makes life worse. However, there are reported cases of people with severe, chronic, life-shortening diseases who greatly value their own lives, and who are thankful for the chance to live as long as possible. At the same time, such cases seem to represent a minority view. The many people who would consider being euthanized when particular circumstances arise—such as severe limited mental capacity due to dementia, being in a permanent vegetative state, or in extremely severe pain (where palliative medicine is no longer effective), or becoming a quadriplegic due to an accident—substantiate this. To live with the knowledge that one may die a terribly unpleasant and painful death is certainly not welcomed by any reasonable person. Therefore, it seems that Peter and Laura should not procreate if there could be a high possibility that their offspring might carry the genetic disease CH (in such a case, they might instead adopt a child as a socially valuable alternative). Alternatively, Peter should be tested and, if the test result is negative, he and Laura may be morally allowed to procreate.

On Some Differences of Analogy between the Right To Know and the Right Not To Know

A related but somewhat different issue is the proponents' claim that the right not to know has the same structure as the right to know, since it is also based in the principle of autonomy, and that therefore the wish of the patient who does not want to receive certain information must *under all circumstances* be respected because it is his or her free and autonomous decision not to be informed. To violate this wish is held to be tantamount to violating the principle of autonomy, which is the foundation of human action. The argument from analogy, however, is misleading because of a misconception of the principle of autonomy in the case of the right to know (see also Macklin 1992 and the Nuffield Report 1993).

John Stuart Mill famously argued that individuals should be free to live their lives according to their own desires, as long as their actions do not harm other people. This fundamental assumption links autonomy with personal liberty, but in such a way as to contain a limiting condition, which can be referred to as the principle of non-maleficence (or "do no harm"). The presence of this limiting condition means that Mill's notion of autonomy is, by nature, *prima facie* and not absolute. If one agrees with this particular way of reasoning, then it follows that one is not always free to act according to one's own desires, precisely because particular actions might harm other people, which would then make those actions, morally speaking, a deal breaker.

If parents like Peter and Laura want to procreate despite their heightened suspicion that they may harm their offspring by passing on a fatal and incurable disease, they cannot morally justify their action by claiming the right not to know and thereby refusing to consent to any available testing. The principle of autonomy, as we have just seen, contains a limiting condition, which is also effective concerning the right not to know. Hence, there seems to be a *parental duty* in some cases either to abstain from procreation (contrary to Asscher and Koops 2010) or to allow for testing, and thereby avoid causing serious harm to any possible offspring who might experience an unpleasant and early death (here, I do not claim that this argument also concerns future children with impairments unless the particular impairments are so severe that it is a so-called "life that is not worth living.")[11] However, if there is a way to test the fetus for the lethal disease while respecting the at-risk parent's wish not to receive any information related to his or her own health, then one can resolve this issue properly (Asscher and Koops 2010). That presupposes, however, that the embryos are tested outside the womb, and then are implanted if they do not carry the CH mutation. However, it would be unrealistic to assume that such an expensive and burdensome procedure will be provided for all possible cases around the world. The more realistic picture is that only a few cases will be treated in this way, whereas in all other such cases the parents will procreate knowing that they may possibly pass on an incurable, lethal disease to their offspring.

One might object, however, that, according to the argument used here that Peter and Laura should not procreate given the heightened risk of passing on a serious disease to their offspring, Peter himself might not have been given the opportunity to live in the first place, if he was the carrier of CH and his parents knew in advance about this possibility. A possible response could emphasize two points: On one hand, Peter's life must be fully respected and protected once he is born, but on the other hand, it may be considered acceptable to abort embryos and fetuses that carry CH while they are unborn. Of course, using the physical event of birth as an ethical dividing line is itself ethically disputed, but this moderating position has proved to be a good pragmatic solution in various situations.

Autonomy-based Conflicts

The right to know is undoubtedly based on the principle of autonomy, and its implementation was meant to promote a person's ability to make well-informed decisions in medical cases. The right to know, however, is not limited to the field of medicine; rather, it is a general feature of ethical reasoning and decision-making. Lack of information concerning important decisions is usually seen (in virtually any situation) as a serious problem, not a blessing.

Proponents of the right not to know argue that this right is based on the very same principle of autonomy as the right to know. The crucial point here is that the lack of some admittedly *substantial* information concerning one's own life path, or that of another person (e.g., one's own offspring) is seen not as a problem but rather as "unnecessary or undesired knowledge" in the course of the pursuit of one's own happiness, "particularly when it concerns information that foreshadows an early and unpleasant death" (Asscher and Koops 2010, 31).

On some occasions, the two supposed autonomy-based rights (the right to know and the right not to know) are in conflict with each other, which obviously poses a problem for the correct understanding of the nature of the principle of autonomy. How can a valid principle justify two mutually exclusive, and in some cases conflicting rights? One might respond that the right not to know is *absolute* by nature and therefore prevails over the *prima facie* right to know. This, however, seems to be an awkward position—besides the fact that it asserts a specific absolute right because it is unclear why the *prima facie* principle of autonomy should determine the nature of both rights in different ways, i.e., that one right is absolute and the other is *prima facie*. It seems, rather, that the right not to know is neither *absolute* nor really (fully) justified by the principle of autonomy itself. Assuming a valid right not to know even in cases that involve causing great actual or potential harm to others is unreasonable and unconvincing.

Conclusions

I strongly believe that one should respect a person's wish not to be informed if he or she indicates this preference. At the same time, I also strongly believe that in many practical cases the so-called right not to know is overridden by other considerations, such as the principle of non-maleficence. The latter principle does not necessarily always rule out the supposed right not to know; ethical judgements certainly depend on the particular case, and cannot be assumed beforehand. I do empathize with parents who have a family history of a serious and perhaps incurable lethal disease, and who want to procreate, but I believe that adhering to the right not to know in such a case undermines the parental duty toward one's possible offspring. "Do no harm"—and hence do not procreate—should be the prevailing principle in all such situations.

Acknowledgements

I would like to thank Mariacarla Gadebusch Bondio and Francesco Spöring for their helpful comments on a previous version of this chapter.

Notes

1 For a brief overview of several positions and arguments in the debate see Chadwick (1997, 17–21).
2 Macklin and the Nuffield Report claim that the physician's duty of confidentiality is not absolute, and that therefore it might be the case in some instances—for example, to serve the public interest because of potential health risks to other people—that the duty of confidentiality could be outweighed by "a legitimate right to know" (see, Husted 1997, 56). However, the Nuffield Report eventually rejects, on pragmatic grounds, a "legally enforceable duty of disclosure to family members placed on people who have been tested and on their medical staff" (Nuffield Council on Bioethics 1993, 53; Husted 1997, 56).
3 An absolute right is a right that cannot be overridden by any other right or duty.
4 Chorea Huntington is a dominant genetic neurodegenerative disease, located on the fourth chromosome, which causes both physical and cognitive deterioration. It is fatal and incurable.
5 Malek and Daar argue that there is a "parental duty" to obtain a preimplantation genetic diagnosis in cases of a history of a serious medical condition linked to a high probability of passing the lethal disease on to the next generation. In their view, it is morally good to promote the well-being of the future child, to expand the possibilities for healthy offspring, and to prevent serious impairments in order to avoid unfair social and economic disadvantages for the particular child.

6 Asscher and Koops (2010) attempt to reconcile the right not to know with a particular way of testing the future generation by appealing to the exclusion test. This solves the problem of not informing the at-risk parent in case he or she has Chorea Huntington while, at the same time, seeking to avoid having offspring who might carry this incurable lethal disease.

7 However, there is a great body of literature on the ethics of PGD, and it ranges from a rather morally neutral (or somewhat positive) stance for the medical benefit of the next generation (Harris 1998) to an absolute refusal to use that technology in order to prevent impairments (Asch 2000). Furthermore, Sandel (2007) has repeatedly claimed that the use of PGD followed by abortion or selection of embryos shows a lack of humility concerning our ability to control the world and a failure to appreciate the giftedness of life. For further references, see Malek and Daar (2012, 4).

8 My translation: "A prenatal genetic test that aims to determine the genetic characteristics of an embryo or a fetus in order to detect a disease, the onset of which will occur only after the 18th year of life according to the generally acknowledged current state of medical science and technology, is not permitted".

9 For the history of informed consent, see Faden and Beauchamp (1986). For some brief historical remarks on the notion of autonomy and informed consent see also my article "Medical Paternalism and Patient Autonomy" (Gordon 2012b).

10 Beauchamp and Childress (2009, 114–15) offer a list of seven major rival standards of incompetence often used by health care professionals to determine the capacity of the particular patient, ranging from weak to strong standards. They place those seven standards in three categories as follows: (1) Stating a preference: inability to express or communicate a preference or choice; (2) Understanding and appreciating one's own situation: inability to understand one's situation and its consequences, inability to understand relevant information; (3) reasoning through a consequential life decision: inability to give a reason, inability to give a rational reason (although some supporting reasons may be given), inability to give a risk/benefit-related reason (although some rational supporting reasons may be given), inability to reach a reasonable decision (as judged according to a reasonable person standard).

11 "The disability rights objection would be highly implausible for those conditions, such as Tay-Sachs or Lesch-Nyhan, which are so bad that it is quite clearly in the interests of the sufferer to cease existing. To say of such lives that they are worth starting or, worse still, that they are not bad at all, would be to strain credulity beyond all reasonable bounds" (Benatar 2006, 114–15).

References

Amundson, R. 2010. "Quality of Life, Disability, and Hedonic Psychology." *Journal for the Theory of Social Behaviour* 40(4):374–92.

Andorno, R. 2004. "The Right Not to Know: An Autonomy Based Approach." *Journal of Medical Ethics* 30(5):435–40.

Asch, A. 2000. "Why I Haven't Changed My Mind about Prenatal Diagnosis: Reflections and Refinements." In *Prenatal Testing and Disability Rights*, edited by E. Parens and A. Asch, 234–58. Washington: Georgetown University Press.

Asscher, E., and B.-J. Koops. 2010. "The Right Not to Know and Preimplantation Genetic Diagnosis for Huntington's Disease." *Journal of Medical Ethics* 36(1):30–3.

Baker, R. 1998. "A Theory of International Bioethics: The Negotiable and Non-Negotiable." *Kennedy Institute of Ethics Journal* 8(3):233–73.

Beauchamp, T.L., and J.F. Childress. 2009. *Principles of Biomedical Ethics*, 6th ed. Oxford: Oxford University Press.

Benatar, D. 2006. *Better Never to Have Been: The Harm of Coming into Existence.* Oxford: Oxford University Press.

Chadwick, R. 1997. "The Philosophy of the Right to Know and the Right Not to Know." In *The Right to Know and the Right Not to Know*, edited by R. Chadwick, M. Levitt and D. Shickle, 13–22. Brookfield: Ashgate.

Chadwick, R., Levitt, M., and D. Shickle, eds. 1997. *The Right to Know and the Right Not to Know.* Brookfield: Ashgate.

Faden, R.R., and T.L. Beauchamp. 1986. *A History and Theory of Informed Consent.* New York: Oxford University Press.

Gordon, J.-S. 2012a. "Bioethics." In *Internet Encyclopedia of Philosophy.* Available at http://www.iep.utm.edu [accessed March 10, 2016].

Gordon, J.-S. 2012b. "Medical Paternalism and Patient Autonomy." In *Medical Ethics*, edited by M. Boylan, 72–83. Hoboken, NJ: Wiley-Blackwell.

Husted, J. 1997. "Autonomy and a Right Not to Know." In *The Right to Know and the Right Not to Know*, edited by R. Chadwick, M. Levitt, and D. Shickle, 55–68. Brookfield: Ashgate.

Knoppers, B.M. 2014. "From the Right to Know to the Right Not to Know." *The Journal of Law, Medicine and Ethics* 42(1):6–10.

Macklin, R. 1992. "Privacy and Control of Genetic Information." In *Gene Mapping*, edited by G.J. Annas and S. Elias, 157–72. Oxford: Oxford University Press.

Malek, J., and J. Daar. 2012. "The Case for a Parental Duty to Use Preimplantation Genetic Diagnosis for Medical Benefit." *American Journal of Bioethics* 12(4):3–11.

Malpas, P. 2005. "Would You Like to Know What Is Wrong with You? On Telling the Truth to Patients with Dementia." *Journal of Medical Ethics* 26(2):108–13.

Ministerium der geistlichen, Unterrichts- und Medizinal-Angelegenheiten. 1901. *Zentralblatt der gesamten Unterrichtsverwaltung in Preussen.* 1900 (Berlin Code).

Nuffield Council on Bioethics. 1993. "Nuffield Report on Genetic Screening: Ethical Issues." Available at http://nuffieldbioethics.org/wp-content/uploads/2014/07/Genetic_screening_report.pdf [accessed March 10, 2016].

Ost, D.E. 1984. "The 'Right' Not to Know." *The Journal of Medicine and Philosophy* 9(3):301–12.

Rehmann-Sutter, C., ed. 2009. *Disclosure Dilemmas: Ethics of Genetic Prognosis after the "Right to Know/Not to Know" Debate.* Farnham: Ashgate.

Sandel, M.J. 2007. *The Case against Perfection: Ethics in the Age of Genetic Engineering.* Cambridge: Harvard University Press.

Sass, H.-M. 1989. *Medizin und Ethik.* Stuttgart: Reclam.

4 Predictive Diagnostic Testing for Late-Onset Neurological Diseases in Asymptomatic Minors

'Do No Harm' and the Value of Knowledge

Heiner Fangerau, Florian Braune, and Christian Lenk

In a 2007 comment on the bioethical questions posed by advances in genetics, the British magazine *The Economist* concluded that "[t]he best test of whether something should be banned is not 'is it distasteful?' but 'is anyone harmed?'" (The Economist 2007). The author focused on the question of harm, rather than taste, because tastes tend to change over time: "Transplanting organs from the dead to the living was regarded by some as immoral, until it started saving lives. So was in vitro fertilization, until it started creating them" (The Economist 2007).[1] Additionally, the ancient principle of 'do no harm' continues to guide normative medical ethics today (Beauchamp and Childress 2001, 116–17). The challenging task therefore is to decide if a scientific innovation can be fundamentally wrong in an ethical sense. Otherwise it might be the ethical principle itself that in specific cases is cited wrongly.

Technologies such as whole-genome sequencing have provided new possibilities for disease prediction (Hawkins et al. 2011). Physicians, geneticists, stakeholders, consumers, patients, and medical ethicists (to name but a few) disagree on whether these technologies offer valuable opportunities for prediction and prognosis or instead cause harm, are dangerous or a curse. Due to advances in genetic research and the increasing number of genetic tests available in the health market, discourse about the risks, opportunities, and ethical implications of genetic testing is ongoing. In particular, the risks posed by predictive genetic testing for incurable diseases have entered the public arena.

Agreement on the helpfulness and usefulness of predictive genetic knowledge of diseases seems to have been reached, in the presence of available therapies for diseases and fatal predispositions identified, with the ability to cure or at least improve prognosis. When no therapy is available, however, predictive genetic knowledge may have problematic consequences, ranging from those related to the dangers of eugenics on a social level to a psychological burden on the level of affected individuals. Fairy tales and ancient myths are full of dark prophecies and their fatal effects, and illustrate the possibly devastating social, emotional, and psychological consequences of ominous predictions.[2] At the same time, predictive knowledge may be seen as useful for the planning of one's life and family, and for the advancement of science in general.

For some paradigmatic genetic disorders, this problematic constellation of conflicting goals has been discussed thoroughly in recent decades. For example, medical societies and patient associations have established procedures for the testing and counselling of persons possibly affected by Huntington's disease (Dufrasne et al. 2011). Whereas Huntington's is a monogenetic disorder with almost complete penetrance (McNeil et al. 1997), genetic constellations without such a high predictive value have been identified for many disorders. For instance, the number of identified genes that increase the likelihood of the development of late-onset neurodegenerative diseases, such as amyotrophic lateral sclerosis (ALS) and Alzheimer's disease (AD), is increasing continually. This situation results in some uncertainty regarding the probability of a genetically predicted disorder occurring (Fryer 2000, 284, Table 4.1). No therapy is currently available for ALS or AD. Because the penetrance of these genes remains unclear, long-term studies in families, including children, are currently under way.[3]

Several genetic associations have published guidelines for predictive testing based on their experience and knowledge. These guidelines reflect and try to achieve a balance among different viewpoints surrounding the concepts of autonomy, the status and 'property rights' of genetic information, truth telling, and the right to know or not to know.[4] A specific problem is posed by the testing of children for disorders for which no therapy exists. The first issue is whether these tests should be permitted or banned. If tests are permitted and available for children, the second issue is that parents and third parties, rather than children, decide whether predictive genetic knowledge would be useful or harmful. One example of the consideration of such issues can be found in the "Guidelines for the Molecular Genetics Predictive Test in Huntington's Disease" (International Huntington Association 1994). More general guidelines addressing the well-being of children in the context of genetic testing, or medical confidentiality with respect to the disclosure of familial genetic information, include the "Recommendations of the European Society of Human Genetics: Genetic testing in asymptomatic minors" (2009) and the advice provided by the American Society of Human Genetics, Social Issues Subcommittee on Familial Disclosure in the "Professional Disclosure of Familial Genetic Information" (1998).

Our discussion probes the issue of whether genetic testing of minors for late-onset neurodegenerative disorders should be recommended. As have others in different contexts (cf. Rhodes 2006), we will examine:

- whether minors should be tested for genetic disorders for which no therapy exists, and
- whether children and their families should be informed about the detection of genetic disorders for which no therapy is available.

Our analysis is based on guidelines for genetic testing published in German and English (European and North American guidelines; Table 4.1), and qualitative interviews on this issue with geneticists, neurologists, and possibly affected persons.[5] After a short overview of the key points brought forward in these guidelines, we discuss these points in relation to some thoughts put forward in the political philosophy of the American philosopher John Rawls (1921–2002). We end with suggestions and a short discussion.

Table 4.1 Selection of Guidelines regarding genetic testing

Guideline and Year of Publication	*Late-onset* **disease addressed**	*Testing of children* **addressed**	*Position on testing concerning children (permissive: "yes" or not permissive "no")*
• *Deutscher Ethikrat, 2013*, Die Zukunft der genetischen Diagnostik—von der Forschung in die klinische Anwendung	• Yes	• Yes	• No
• *EFNS Guidelines, European Journal of Neurology 2011, 18: 207–217*, EFNS guidelines for the molecular diagnosis of neurogenetic disorders: motoneuron, peripheral nerve, and muscle disorders	• Yes	• No	• —
• *Gendiagnostik-Kommission am Robert Koch-Institut, 2011*, Richtlinie der Gendiagnostik-Kommission (GEKO) zu genetischen Untersuchungen bei nicht-einwilligungsfähigen Personen nach § 14 in Verbindung mit § 23 Abs. 2 Nr. 1c GenDG	• No	• No	• —
• *Deutsche Gesellschaft für Humangenetik e.V. (GfH), Berufsverband Deutscher Humangenetiker e.V. (BVDH), 2011*, S2-Leitlinie, Humangenetische Diagnostik und genetische Beratung	• Yes	• Yes	• Yes
• *European Society of Human Genetics, 2009*, Genetic testing in asymptomatic minors: recommendations of the European Society of Human Genetics	• Yes	• Yes	• No

(*Continued*)

Table 4.1 (Continued)

Guideline and Year of Publication	*Late-onset* **disease addressed**	*Testing of children* **addressed**	*Position on testing concerning children (permissive: "yes" or not permissive "no")*
• *Borry, 2009*, Genetic testing in asymptomatic minors, Background considerations towards ESHG Recommendations	• Yes	• Yes	• No
• *AWMF, Leitlinien der Deutschen Gesellschaft für Neurologie, 2008*, Genetische Diagnostik und Beratung bei neurologischen Erkrankungen	• Yes	• Yes	• No
• *European Commission 2004*, 25 Recommendations on the ethical, legal, and social implications of genetic testing	• No	• Yes	• —
• *Canadian Paediatric Society, 2003*, Guidelines for genetic testing of healthy children	• Yes	• Yes	• No
• *Bundesärztekammer, 2003*, Richtlinien zur prädiktiven genetischen Diagnostik	• Yes	• Yes	• No
• *American Academy of Pediatrics, 2001, published in "Pediatrics"*, Ethical Issues with Genetic Testing in Pediatrics	• Yes	• Yes	• No
• *ASHG Statement, 1998*, Professional Disclosure of Familial Genetic Information	• No	• No	• —
• *J Med Genet, 1994*, Guidelines for the molecular genetics predictive test in Huntington´s disease	• Yes	• No	• —

Genetic Testing of Children: Inappropriate?

Since 1994, several societies and associations have issued guidelines and recommendations on genetic testing. The sequencing of the human genome in 2001 seems to have triggered the publication of these guidelines. A 2001 British survey of public opinion on pre-implantation genetic diagnostic testing conducted by the Human Fertilisation and Embryology Authority and the Human Genetics Commission represents a first initiative to frame the access to genetic testing. A majority of respondents approved the approach suggested in the consultation document that a late-onset disorder should be among the factors informing consideration of such testing (HFEA and HGC 2001).

Many guidelines address late-onset disorders and the testing of children explicitly, and several of the critical issues put forward can be easily identified as general questions about the prediction of such diseases. These questions include: Are genetically affected children without symptoms ill? To what extent is the concept of a 'healthy ill' child reasonable or useful? To what extent do therapeutic options influence the decision to test? Is scientific research on children who will suffer from these disorders adequate without immediate prospects of therapy?

Although medical journals, policy statements, and guidelines contain several arguments for the testing of children for genetic disorders, such as the right to know (Rhodes 2006), the value of knowledge as being in the best medical interest of the child,[6] the earliest use of possible treatment options as they become available (e.g., Kent 2005), and support for the child before symptoms appear (Harris et al. 2012), most guidelines take a critical perspective on the testing of minors.[7] The authors of such materials generally argue for better communication of risk (e.g., ESHG 2009), and demand the appraisal and inclusion of minors when testing is performed because treatment options or effective support options exist (Borry et al. 2009, 712). Many guidelines address concerns about the deleterious emotional and psycho-social consequences of testing for children, who lack the intellectual capacity to understand the information revealed (Borry et al. 2009, 714). These consequences include the danger of a child's acceptance of the role of a sick person, imposed initially by his or her parents; the dangers of stigmatization and discrimination; and the overriding of the child's wishes by fearful parents.[8]

Overall, these issues reflect the benefits and dangers of paternalism. The guidelines tend to recommend against the testing of children. However, this standpoint is challenged by the reality that genetic tests are readily available on the health market. The real task seems to be the recognition of further tests in the pipeline with an even broader analytical scope, focusing, for example, on the detection of predispositions in the whole genome—in particular raising questions of informed consent (Greely 2011). This is probably all the more true for sales of genetic tests directly to consumers. Genetic testing has created a dynamic, expanding field, and patients as well as doctors are increasingly urged to make decisions on whether to utilize it.

The divergence of views expressed in the guidelines of similar professional organizations can be explained by the dependence of every fear, hope, and recommendation on specific conditions framing each testing situation. The most important factors (Clayton et al. 2014) seem to be (a) the predictive value of the prognostic test and (b) the epistemic context of the production of the predictive marker (including hope for the development of a therapy). For example, pre-symptomatic and predictive testing are distinct. Pre-symptomatic testing helps to diagnose in advance a disease that will inevitably occur, whereas predictive testing detects genetic susceptibilities that increase the likelihood of disease development in the future. Although pre-symptomatic testing may appear to be less problematic upon initial consideration, as it does not trouble consulters about their future without reason, the example of the monogenetic Huntington's disease test reveals the complexity of the issue. The age of onset of this disease, which is very important for the planning of one's life after diagnosis, differs immensely, minimizing the potential benefit of the predictive knowledge and perhaps resulting in constant apprehension as one awaits onset.

The context of the testing of minors also influences ethical inferences. Testing for research might be viewed differently from testing on demand from parents, or children themselves. Additionally, the perceived role or involvement of the parents, other family members, or the professional understanding of the role of physicians in the testing process might lead to different conclusions. As an example, we would point out the differences in recommendations issued by the American Academy of Pediatrics (AAP) and the American College of Medical Genetics (ACMG) regarding the genetic testing of children. Clayton et al. showed that although the ACMG was involved in the development of both recommendations, the conclusions regarding testing differed (Clayton et al. 2014). Whereas the guideline issued together with the pediatricians concluded that physicians should abstain from testing when treatment options do not exist, the guideline published solely by the ACMG argued for examining a number of variants that may lead to interventions one day and recommended that results for 56 genes from whole-genome sequences should be revealed. Integral to this discrepancy are the different evaluations of the roles and intentions of participants in the decision-making process, which differ according to the testing method.

Whole-genome sequencing is currently viewed as holding the most potential for future genetic testing. As this new technology appears to be creating a framework for a new 'ethical reality', re-examination of the ethical basis for the careful, protective position against the genetic testing of minors for late-onset diseases is warranted. This line of reasoning is based on concerns about the misuse of information and possible negative consequences for the child.

The critical view of genetic testing of children primarily addresses the possibility of negative psycho-social effects on the child. However, a particular opinion included in one set of guidelines explicitly mentions another issue, which is only implicit in other guidelines: the fear that parents or society in general might withdraw resources from affected children to avoid 'wasting' them on individuals with potentially poor futures.[9] Genetic knowledge thus has moral implications for discourses about fairness, equity, and justice.

How Can We Safeguard Minors?

We think about the protection of basic rights and the distribution of possible advantages and disadvantages as a result of asymptomatic testing in the case of minors. In this context, we want to point out how John Rawls addressed these aspects as criteria for a just society in his main work, the "Theory of Justice." We want to introduce these criteria into the present context for an estimation of the risks and benefits which may occur for minors after asymptomatic testing in society. Rawls in this theory developed the idea that one understanding of justice could be equated to fairness in a society with members holding basically the same rights. He ideally defined society as a closed circle, in which structural decisions determine the lives of its members. The members again cooperate according to their capabilities, but cooperative goods should be distributed equally (Wenar 2013 online). Along these lines Rawls attempted to formulate principles of justice using a thought experiment. In this experiment, people must move from a fictional 'original position' to principles of justice without knowing their future positions in society, as they are blinded by an imaginary veil of ignorance. Rawls assumed that everyone involved in this scenario would prefer maximum liberty and equality, to reduce the risk of being placed in a disadvantageous position in society.

Rawls defined two rules that must be obeyed to create a just society (1971, 13):

> The first requires equality in the assignment of basic rights and duties, while the second holds that social and economic inequalities, for example inequalities of wealth and authority, are just only if they result in compensating benefits for everyone, and in particular for the least advantaged members of society.

The question now is, whether the testing of asymptomatic minors might harm their basic rights and duties and whether there are any "compensating benefits" if the testing results in "social and economic inequalities." Regarding their rights, one might say that testing asymptomatic minors might fail to respect their future right of autonomy. They are especially vulnerable regarding possible risks of being overruled in their choices. Additionally, they are in danger of experiencing harm in their development by for example being excluded from resources. If children with genetic preconditions are seen therefore as least advantaged, Rawls' rules should result in compensating benefits for them, such as therapy, counselling, or other forms of support. If these benefits are not available due to lack of resources or uncertainties regarding benefit-risk rationing, we conclude that the protection of these children—in other words, abstaining from testing—should be the first priority. For late-onset diseases, we recommend the postponement of testing until a minor is able to decide for him- or herself, since compensating resources cannot be guaranteed.

Arguments such as ownership of genetic information by entire families, rather than possibly affected minors alone, can also be countered by the principle of protecting the 'least advantaged'. Thus, we concur with the careful perspective taken in most guidelines that the 'right not to know' is in the best interests of the child. This approach avoids stigmatization and the negative effects of alarming predictions on a child's development, and does not affect care for children in competent families.

Do we, however, have the political right to interfere with families' decisions and approaches to caring for children? The answer to this question centers on the consideration of harm to the child. Possible forms of harm are discrimination, anxiety about the future, parental overprotection, and depression due to a fatal prognosis. These risks could be assessed properly in long-term cohort studies involving the testing of children, disclosure of results to families, and follow-up to monitor the children's well-being. Such research, however, would expose children to potential harm with less justification than in the clinical context of genetic testing.

Coming back to our introduction and *The Economist*'s plea to ask in ethical evaluations of new technologies whether anyone is harmed, we have to state that the testing of asymptomatic minors entails possible hazards for them on various levels. For this reason, we again recommend that children not be tested for late-onset diseases for which no therapy is available.

On the basis of these considerations, we concur with the 'careful' approach presented in relevant guidelines:

- Because children are vulnerable, testing should, if performed at all, be accompanied and followed by 'compensating benefits'.
- In the absence of 'compensating benefits' or clarity regarding the parties responsible for compensation (e.g., government, parents, insurance), we recommend protection of children from potentially harmful information.
- Overall, we strongly recommend the postponement of genetic testing until children can decide for themselves.

Notes

1 Journalists writing for *The Economist* usually remain anonymous.

2 A good example is that of the fairy tale "Sleeping Beauty". After a dark prediction for the future of their daughter, the king and queen try to protect her from this danger, but ultimately fail. The daughter is unhappy with what she perceives as 'overprotection'.

3 Cf. Mand et al. (2012): "In keeping with the overall lack of systematic work in the area, commentators have rarely discussed which of their arguments specifically apply to infants and younger children, older children or adolescents. [. . .] the third possibility is to systematically collect empirical evidence to test the claims regarding negative consequences and reform guidelines based on this evidence. We advocate the latter, which would allow the systematic collection of evidence in controlled and clearly defined settings". See also Hens et al. (2011).

4 On the right to know or not to know, see also Lenk and Frommeld (2015).

5 The interviews mentioned above were conducted in the context of a research project at Ulm University regarding predictive genetic testing in minors for

adult-onset genetic diseases. The project was funded from 2013 to 2015 by the German Federal Ministry of Education and Research.

6 Cf. Recommendations of the European Society of Human Genetics (ESHG 2009, 721): "(3) The parents (or legal guardians) should participate as much as possible in the decision-making process regarding the health care of their children. If the decision of the minor's parents or legal guardians is not to the direct benefit of the minor, health care professionals have the responsibility to defend the interests of the minor".

More direct: "When talking about health-care decisions involving children, the concept of 'best interests' takes a more central position than the concept of informed consent" (Borry et al. 2009, 712).

7 An overview of these arguments from a diachronic perspective and further references are offered by Mand et al. (2012, 521 Box 1).

8 Cf. Deutscher Ethikrat (2013, 158): "The use of genetic diagnostics during pregnancy would enable parents to bind the realization of the desire to have children to self-imposed conditions which the child must meet. This way an ethically doubtful attitude of parents towards their future children may be developed. The children would not be regarded anymore as equal subjects, who are to be respected for their own sake, i.e. in their individual suchness. The development of modern reproductive medicine and the broad employment of prenatal diagnostics would instead result in children becoming objects of parental desires and preferences." Translation by the authors – original quote: "*Der Einsatz der genetischen Diagnostik während der Schwangerschaft ermögliche es den Eltern, die Realisierung des Kinderwunsches an selbst gesetzte Bedingungen zu binden, denen das Kind genügen müsse. Auf diese Weise könne sich eine in ethischer Hinsicht fragwürdige Einstellung der Eltern gegenüber ihren künftigen Kindern ausbilden. Diese würden nicht mehr als ebenbürtige Subjekte betrachtet, die um ihrer selbst willen, das heißt in ihrem individuellen Sosein zu achten seien. Die Entwicklung der modernen Reproduktionsmedizin und der breite Einsatz pränataler Diagnostik führten vielmehr dazu, dass Kinder in immer stärkerem Maß als Objekte elterlicher Wünsche und Präferenzen in den Blick gerieten*".

9 On this notion, cf. Committee on Bioethics (2001, 1454): "Further, testing in childhood inappropriately eliminates the possibility of future autonomous choice by the person and risks stigma and discrimination. Unless there is anticipated benefit to the child, pediatricians should decline requests from parents or guardians to obtain predispositional genetic testing until the child has the capacity to make the choice", and Borry et al. (2008): "It is difficult for the child to live in the knowledge that, while not sick now, he or she has a greater likelihood of becoming sick in the future".

References

American Society of Human Genetics (ASHG) Social Issues Subcommittee on Familial Disclosure. 1998. "Professional Disclosure of Familial Genetic Information." *American Journal of Human Genetics* 62:474–83.

Anonymous. 2007. "Clones to the Right of Me, Jokers to the Left: The Study of Genetic Chimeras Should Be Encouraged, Not Banned." *The Economist*, print edition, January 13, 2007. Available at http://www.economist.com/node/8527437 [accessed May 25, 2015].

Beauchamp, T. L., and J. F. Childress. 2001. *Principles of Biomedical Ethics*. New York: Oxford University Press.

Borry, P., Evers-Kiebooms, G., Cornel, M. C., Clarke, A., and K. Dierickx. 2009. "Genetic Testing in Asymptomatic Minors: Background Considerations towards ESHG Recommendations." *European Journal of Human Genetics* 17:711–19.

Borry, P., Goffin, T., Nys, H., and K. Dierickx. 2008. "Predictive Genetic Testing in Minors for Adult-Onset Genetic Diseases." *Mount Sinai Journal of Medicine* 75(3):287–96.

Clayton, E.W., McCullough, L.B., Biesecker, L.G., Joffed, S., Friedman Rosse, L., and S.M. Wolff. 2014. "Addressing the Ethical Challenges in Genetic Testing and Sequencing of Children." *The American Journal of Bioethics* 14(3):3–9.

Committee on Bioethics. 2001. "Ethical Issues with Genetic Testing in Pediatrics." *Pediatrics* 107(6):1451–4.

Deutscher Ethikrat. 2013. "Die Zukunft der Genetischen Diagnostik—Von der Forschung in die Klinische Anwendung." Available at http://www.ethikrat.org/dateien/pdf/stellungnahme-zukunft-der-genetischen-diagnostik.pdf [accessed May 25, 2015].

Dufrasne, S., Roy, M., Galvez, M., and D.S. Rosenblatt. 2011. "Experience over Fifteen Years with a Protocol for Predictive Testing for Huntington Disease." *Molecular Genetics and Metabolism* 102(4):494–504.

European Society of Human Genetics (ESHG). 2009. "Genetic Testing in Asymptomatic Minors." *European Journal of Human Genetics* 17:720–1.

Fryer, A. 2000. "Inappropriate Genetic Testing of Children." *Archives of Disease in Childhood* 83:283–5.

Greely, H.T. 2011. "Get Ready for the Flood of Fetal Gene Screening." *Nature* 469:289–91.

Harris, E.D., Ziniel, S. I., Amatruda, J.G., Clinton, C.M., Savage, S.K., Taylor, J.D., . . . I.A. Holm. 2012. "The Beliefs, Motivations, and Expectations of Parents Who Have Enrolled Their Children in a Genetic Biorepository." *Genetics in Medicine* 14(3):330–7.

Hawkins, A.K., Ho, A., and M.R. Hayden. 2011. "Lessons from Predictive Testing for Huntington Disease: 25 Years On." *Journal of Medical Genetics* 48(10):649–50.

Hens, K., Lévesque, E., and K. Dierickx. 2011. "Children and Biobanks: A Review of the Ethical and Legal Discussion." *Human Genetics* 130:403–13.

Human Fertilisation Embryology Association (HFEA) and Human Genetics Commission (HGC). 2001. "Outcome of the Public Consultation on Preimplantation Genetic Diagnosis." Available at http://www.hfea.gov.uk/cps/rde/xbcr/hfea/PGD_outcome.pdf [accessed May 25, 2015].

International Huntington Association and the World Federation of Neurology Research Group on Huntington's Chorea. 1994. "Guidelines for the Molecular Genetics Predictive Test in Huntington's Disease." *Journal of Medical Genetics* 31(7):555–9.

Kent, A. 2005. "We Can Change the Future." *EMBO—Reports* 6(9):801–4.

Lenk, C., and D. Frommeld. 2015. "Different Concepts and Models of Information for Family-Relevant Genetic Findings: Comparison and Ethical Analysis." *Medicine, Health Care and Philosophy* 18(3):393–408.

Mand, C., Gillam, L., Delatycki, M.B., and R.E. Duncan. 2012. "Predictive Genetic Testing in Minors for Late-Onset Conditions: A Chronological and Analytical Review of the Ethical Arguments." *Journal of Medical Ethics* 38:519–24.

McNeil, S.M., Novelletto, A., Srinidhi, J., Barnes, G., Kornbluth, I., Altherr, M.R., . . . R.H. Myers. 1997. "Reduced Penetrance of the Huntington's Disease Mutation." *Human Molecular Genetics* 6(5):775–9.

Rawls, J. 1971. *A Theory of Justice*. Cambridge: Harvard University Press.

Rhodes, R. 2006. "Why Test Children for Adult-Onset Genetic Diseases?" *Mount Sinai Journal of Medicine* 73(3): 609–16.

Wenar, L., "John Rawls," *The Stanford Encyclopedia of Philosophy*, Winter 2013 ed., edited by Edward N. Zalta. Available at http://plato.stanford.edu/archives/win2013/entries/rawls/

5 Incidental Findings in Genetic Testing

Elke Holinski-Feder and
Verena Steinke-Lange

Summary

The great technological advances seen in recent years in the area of genetic testing have significantly changed the way we approach genetic analysis. Until recently, genetic diagnostics have been an expensive and time-consuming measure which, in everyday clinical practice, has been performed only reservedly, and in cases of strong clinical suspicion of a specific inherited disease. As a result, diagnostic options were decidedly limited. Atypical clinical phenotypes were often not correlated with a specific disorder when comprehensive genetic testing would have enabled accurate diagnosis. In order to save money, clinical phenotypes for which mutations in numerous genes have been described in the literature normally met with analyses for only the most common genetic mutations. For many families, the causative mutation was therefore not identified, and relatives were correspondingly denied the advantages of predictive testing.

With the advent of next-generation sequencing (NGS) technology, it is now possible to analyze a variety of genes simultaneously, with relatively limited expenditure of money and time. That is why this method has increasingly become a part of routine genetic diagnostics in recent years, even in places where the indications for its use, and the coverage of its costs by local health care systems, have been inconclusive.

Independent of its clinical advantages for patients, the broad, simultaneous analysis of multiple genes nevertheless produces some problems. The analytical technique normally targets not only genes associated with the particular disease in question, but also genes associated with other unrelated genetic disorders. NGS analysis therefore often produces incidental findings in genes not necessarily meant to be analyzed. There are as yet no uniform guidelines for handling incidental findings in clinical genetics—no guidelines which would, for example, regulate which results would be given to the patient or prescribe the form of his or her consent. Dealing with incidental findings thus presents a great challenge to those performing the analyses, as well as to physicians and patient counselors. At present, professional associations concerned with genetic testing are attempting to create corresponding guidelines. The debate within these associations, as well as within ethics organizations and the general population, is ongoing. The following article serves as an overview of the genetic basis of incidental findings in clinical genetics, and of the possible approaches to handling them, thereby contributing to the ongoing debate.

Basic Genetic Principles

The human genome is comprised of an approximately 1.8 meter long strand of DNA made up of some three billion individual base pairs distributed across twenty-three pairs of chromosomes. Only a small portion of this (approx. 3–4%) consists of genes (i.e., genomic regions), which are read by enzymes and, for example, converted into proteins. The number of genes is currently estimated to be about 30,000, with each gene encompassing anywhere from 200 to 10,000 base pairs. Most genes are present in two copies, one inherited through the maternal, and one through the paternal chromosome in every chromosomal pair. To date, over 5,000 monogenetic diseases caused by individual mutations in single genes have been identified. Depending on the pattern of inheritance, a disease occurs when only one of the gene copies is affected (dominant pattern of inheritance), or only when both the maternal and paternal copies carry the corresponding mutation (recessive pattern of inheritance).

Genetic Testing

Clinical suspicion of a particular inherited disease in a patient may be confirmed by testing for, and identification of, known causative mutations in the genes associated with that disease. In this context, analysis is referred to as **genetic diagnostics**, since the purpose of running the test is to identify the cause of—and thus diagnose—the disorder. For well-defined clinical phenotypes such as Neurofibromatosis Type 1, for example, only very few causative genes come into question. In these cases, one need order only a targeted analysis of the relevant genes. With non-specific disorders such as epilepsy, or developmental delay in childhood, clinical delineation of a specific genetic cause is usually difficult; diagnostic clarification is most often made with tiered testing, or, more recently, with a growing number of gene panel analyses (see below).

Predictive genetic testing refers to the identification of a genetic mutation in a healthy individual in order to assess that person's risk of developing an inherited disorder in the future. These analyses are performed in cases where an inherited disease, such as hereditary breast and ovarian cancer, for example, has already been confirmed and family members can be tested for a known familial genetic mutation.

The purpose of **testing for carrier status** (or carrier testing), on the other hand, is to determine if the presence of a genetic mutation in an individual represents a risk for that person's offspring of developing a specific genetic disorder. The person tested is normally healthy, and will not develop symptoms of the disease despite being a carrier of a genetic mutation.

Generating Incidental Findings

So far, the gold standard for the genetic analysis of individual genes has been Sanger sequencing, whereby a single sequence is capable of capturing only about 250 to 600 base pairs. In recent years, genetic analysis using **Next-Generation Sequencing** (NGS) has increasingly found its way into clinical use. This test method enables the simultaneous analysis of, for example, a group of genes associated with the same disorder, or with clinically similar genetic disorders by using a so-called gene panel. This saves considerable time and money in comparison to single-gene analysis. In cases with a particularly non-specific clinical phenotype, exome sequencing can be used in conjunction with clinical data to analyze the coding regions of all known disease-related genes simultaneously. Sequencing of the entire genome has already been performed in research settings, but has played as yet no role in day-to-day clinical practice.

With the increasing amount of information gathered by such comprehensive genetic diagnostics comes the problem of interpreting this data. In recent years, thanks to large-scale sequencing projects such as the 1000 Genomes Project, we have learned a great deal about the "normal" variability of the human genome (The 1000 Genome Consortium 2012). However, evaluating a rare genetic abnormality found in an isolated case, and assessing its status as a normal variant or a disease-causing mutation, is in many instances still difficult. At present, every analysis of the entire genome identifies approximately twenty rare, functionally important variants in disease-related genes, the causative function and clinical relevance of which remain unclear.

Incidental Findings in Clinical Genetics

In genetics, mutations identified in the course of testing which are unrelated to the actual clinical phenotype in question but nevertheless significant as regards the health of the tested individual and his or her family are referred to as incidental or additional findings. The larger the genomic region analyzed, the higher the probability of finding such variants.

Among these variants are **causative mutations for known monogenetic disorders**, which at the time of testing have not yet manifested (e.g., a late-onset limb-girdle muscular dystrophy) or been diagnosed. Some of these diseases have effective treatments, such that the identification of causative mutations can be of clinical importance to the tested individual. In cases of familial hypercholesterolemia, for example, timely treatment with cholesterol-lowering medications can significantly reduce the risk of coronary heart disease. With many forms of inherited cancer, close compliance with the appropriate surveillance programs can greatly reduce the risk of advanced tumor development. Clinical relevance is often a criterion for reporting incidental genetic findings.

Furthermore, genetic diagnostics may identify **carrier status for an inherited disease**, which may have severe consequences for future offspring. In cases of autosomal recessive inherited diseases, offspring are only at risk of developing the disease when the other parent is also a carrier of the same genetic mutation, and the abnormal gene copy is inherited from both parents. As most autosomal recessive disorders are rare diseases, this is not very likely. An exception to this rule is, for example, mucoviscidosis (or cystic fibrosis), of which approximately one in every twenty-five people in the European population is a carrier. X-chromosomal recessive disorders (such as Duchenne muscular dystrophy), or autosomal dominant diseases with reduced penetrance (such as DiGeorge syndrome) are associated with a risk of 25% or more when only one parent is a carrier. Incidental identification of carrier status in the course of genetic diagnostics can therefore be of great importance to the tested individual as regards family planning. In some situations, identification of a genetic abnormality in one person may be of direct clinical relevance for family members, even without it being of relevance for the tested individual. An example of this is when a causative mutation for inherited breast and ovarian cancer is found in a man for whom this result represents no risk of disease but which may be associated with a significantly increased risk of breast and ovarian cancer for his female relatives, who can then undergo early detection screening and, if appropriate, undertake prophylactic measures to fight the disease.

Further incidental findings in clinical genetics are variants described in association with an **increased risk of one or more multifactorial diseases.** However, these diseases occur as a result of interaction between various genetic abnormalities and external factors; a concrete risk assessment based on the identification of genetic risk factors alone is normally not possible. Identification of the HLA-B27 antigen, for example, is associated *inter alia* with an increased risk of developing Bechterew's disease, but the exact risk cannot be determined.

Managing Incidental Findings

In the last few years, the growing problem of incidental findings in clinical genetics has triggered a major debate at various levels. A number of professional associations have created guidelines which address the matter of dealing with these findings, some in terms of clinical practice, and some in a scientific context. As befits the subject, human genetic and bioethics organizations in particular have been addressing these issues. This article has taken into consideration the opinions of the most important professional organizations in Europe, North America, and Australia. Some of these, such as the American College of Medical Genetics (ACMG) and the European Society of Human Genetics (ESGH), have proposed guidelines for managing incidental genetic findings which attempt to answer the following questions (Shkedi-Rafid et al. 2014).

Which Incidental Findings Should Be Reported?

When it comes to reporting incidental genetic findings that have direct clinical relevance for the tested individual because they correspond to an elevated risk of developing a treatable disease, and reporting them would therefore enable that person to avoid serious medical consequences, there is no dispute. Many guidelines are in favor of reporting such incidental findings (Berg et al. 2011; Green et al. 2013). There is, however, no consensus when it comes to what constitutes clinical relevance. Representatives of the ACMG have created a list of 57 genes (first and foremost genes associated with inherited cancer syndromes), which they regard as clinically relevant, and mutations which they feel should be actively sought and reported as long as the individual tested has no objections (Green et al. 2013).

However, it is difficult to prove direct clinical relevance for the majority of incidental genetic findings. For example, it is unclear whether or not variants that correspond to a risk for family members or for future offspring should be reported, since they have no direct medical advantage for the person tested. Notwithstanding, knowledge of carrier status for a severe inherited disorder that could manifest in future offspring is regarded by many as clinically relevant, and important information in terms of family planning. Among such disorders are autosomal recessive diseases, such as mucoviscidosis, which can have severe consequences for affected children. Approximately one in every twenty-five people in the general European population is a heterozygous carrier of a causative mutation in the corresponding gene *CFTR*. However, the disease manifests only when both parents are carriers and the causative mutation is inherited from both parents. Similarly, carrier status for hemochromatosis may be discovered (every eleventh person in the general European population is a carrier); only a small percentage of homozygous carriers develop iron deposits in certain tissues and suffer corresponding symptoms (exhaustion, joint pain, Diabetes mellitus, arrhythmia). Another example are X-chromosomal inherited diseases such as Duchenne muscular dystrophy, whereby heterozygous female carriers are often asymptomatic, but their male offspring have a 50% chance of developing the disease. Incidental genetic findings may therefore reveal carrier status for inheritable diseases that are very diverse with regard to both severity and risk factor, such that it is difficult to define a uniform approach to managing these findings. In any event, how to handle incidental findings needs to be clarified in the course of patient counseling (see below).

The practice of not reporting genetic variants for which the causative effect is unknown at the time of evaluation and a clear risk assessment cannot be made has been relatively consistent. Whether or not reevaluation of a variant should be carried out at a later date, and the manner in which its reclassification as, for example, a disease-causing mutation should be reported, is still unclear. Guidelines need to be drawn up which would regulate the approach to managing these variants (van El et al. 2013).

What Form of Patient Information Is Necessary Prior to Genetic Testing?

Fundamentally, every tested person needs to be informed of all possible findings and their consequences prior to testing, so that the tested individual can give his or her informed consent. Given the ever-widening spectrum of genetic analyses made possible by next-generation sequencing and the abundance of possible test results, truly comprehensive patient information is, generally speaking, not possible. Such clarification would also overtax the tested individual, and quite possibly prevent his or her consent to meaningful genetic diagnostics. It is difficult to strike a balance in this regard. There is some agreement, however, that the tested individual should be made aware of the possibility of incidental results, and that the handling of such results should be addressed prior to testing. The Association of Genetic Nurses and Counsellors (AGNC) guidelines provide patients with the option of choosing whether or not they would like incidental findings to be reported. Should this not be the case, any evaluation of incidental findings needs to be limited to those genomic regions directly related to the clinical disorder (Middleton et al. 2014). This approach is also supported by the Royal College of Pathologists of Australasia (RCPA, *Massively Parallel Sequencing Implementation Guidelines*).

What Form of Patient Consent Makes Sense?

Expert opinions vary greatly when it comes to the matter of a patient's consent to the reporting of incidental genetic findings. While within the context of scientific research consent often includes a statement indicating that incidental results unrelated to the research will not be reported, patients in a clinical setting are often presented with the option as to whether or not they would like to be informed of results that may be of clinical importance to them or to their family members. Some go even further: Baylor College, for example, makes consent to the reporting of incidental findings a precondition for genetic analysis, with the justification that the discovery of rare incidental results in the field of medicine is not a specific test to which a person can give his or her consent, but is an inextricable part of comprehensive medical diagnostics. Whether or not the clinical importance of a genetic analysis outweighs the burden of any potential incidental results is thus left to the patient's discretion. Those performing the analyses are thus spared the ethical dilemma of deciding whether or not they are allowed to report incidental genetic findings with important clinical implications for the tested individual. ESHG guidelines go so far as to suggest reporting incidental results even in cases where consent was not given, if reporting them means treatment or prevention of a disease (van El et al. 2013). A similar approach is supported by the Public Health Genomics (PHG) Foundation (2013).

The problem, as well as the difficulties faced by those trying to solve it, can perhaps be elucidated by the following example: Genetic testing is performed for a patient suffering from severely impaired vision. In addition to the genetic diagnosis of an inherited eye disease, a genetic mutation is found which is known to cause colorectal cancer; the patient has a 60% lifetime risk of developing the disease. The patient, already overwhelmed by increasing blindness, refused to give his consent to the reporting of incidental findings. His doctor is now faced with the dilemma of wanting to inform the patient of these findings, and thus of his cancer risk, so that the patient may take appropriate preventative measures such as regular colonoscopies. The arguments on both sides are entirely valid, such that it is difficult to find a general solution to the problem. This is made all the more difficult by the fact that "treatment" can look very different in clinical practice (e.g., daily medications for lowering hereditary hypercholesterolemia versus annual colonoscopies to monitor precancerous or early stages of cancer in cases of hereditary colorectal cancer).

How Can We Limit the Number of Incidental Genetic Findings?

Even though current technologies enable us to glean a wide range of genetic information with a single analysis, this type of analysis is by no means necessary in all cases. Due to the immense effort involved in interpreting the flood of data, this sort of testing is often not even wanted. That is why day-to-day clinical practice usually focuses on limiting diagnostics to those genes related—even if only remotely related—to the clinical diagnosis. These genes are normally grouped in panels according to phenotype, thereby reducing the number of incidental findings. This approach has been recommended by, for example, the ACMG (Green et al. 2013), ESHG (van El et al. 2013), and RCPA (*Massively Parallel Sequencing Implementation Guidelines*). Critics argue that this approach would exclude any additional value for the patient with regard to his or her risk of developing a treatable disease unrelated to the clinical disorder for which he or she was tested. In order to minimize the loss of useful information, guidelines therefore currently favor offering tested individuals the option of being presented with incidental findings that may be of clinical relevance (see above). As described above, there are nevertheless no uniform guidelines to date concerning the kind of incidental findings that would fall into this category, or if known, clinically important mutations should be sought out regardless of the actual clinical question. Since incidental genetic findings may pertain to a wide variety of diseases (mild or severe, treatable or untreatable), there are no standard means (e.g., test quality criteria, disease penetrance) by which one could simply decide which genetic variants are clinically relevant and which are not.

Should Testing Be Performed Differently for Individuals Incapable of Giving Consent?

Some recommending bodies such as the ESHG and AGNC suggest that incidental genetic findings, especially carrier status for diseases that manifest only in adulthood, be handled differently for children than for adults (van El et al. 2013; Middleton et al. 2014). The guidelines of the ACMG, however, do not discriminate in this matter, since such incidental findings could be of clinical relevance to a child's family members, and reporting them would be, by extension, in the child's best interest (Green et al. 2013).

Summary

The number of incidental findings in clinical genetics has grown immensely in recent years due to technical advances in the field of genetic diagnostics. Since it is only a matter of time before whole genome sequencing becomes part of routine testing, and because whole genome sequencing will produce clinically relevant incidental findings in approximately 1% of all cases, it is important to reach a consensus concerning the management of clinically relevant incidental findings. The guidelines created thus far agree with each other on only a few points. Many questions remain unanswered, and the debate is likely to continue for some time.

References

The 1000 Genome Consortium. 2012. "An Integrated Map of Genetic Variation from 1,092 Human Genomes." *Nature* 491:56–65.

Berg, J.S., Khoury, M.J., and J.P. Evans. 2011. "Deploying Whole Genome Sequencing in Clinical Practice and Public Health: Meeting the Challenge One Bin at a Time." *Genetics in Medicine* 13:499–504.

Green, R.C., Berg, J.S., Grody, W.W., Kalia, S.S., Korf, B.R., Martin, C.L., McGuire, A.L., Nussbaum, R. L., O'Daniel, J.M., Ormond, K.E., Rehm, H.L., Watson, M.S., Williams, M.S., and L.G. Biesecker. 2013. "ACMG Recommendations for Reporting of Incidental Findings in Clinical Exome and Genome Sequencing." *Genetics in Medicine* 15:565–74.

Middleton, A., Patch, C., Wiggins, J., Barnes, K., Crawford, G., Benjamin, C., and A. Bruce. 2014. "Position Statement on Opportunistic Genomic Screening from the Association of Genetic Nurses and Counsellors (UK and Ireland)." *European Journal of Human Genetics* 22:955–6.

PHG Foundation. 2013. "Managing Incidental and Pertinent Findings from WGS in the 100,000 Genomes Project." Available at http://www.phgfoundation.org/file/13772/ [accessed March 7, 2016].

The Royal College of Pathologists Australasia (RCPA). "Massively Parallel Sequencing Implementation Guidelines." Available at http://www.rcpa.edu.au/getattachment/7d264a73–938f-45b5–912f-272872661aaa/Massively-Parallel-Sequencing-Implementation.aspx [accessed March 7, 2016].

Shkedi-Rafid, S., Dheensa, S., Crawford, G., Fenwick, A., and A. Lucassen. 2014. "Defining and Managing Incidental Findings in Genetic and Genomic Practice." *Journal of Medical Genetics* 51:715–23.

van El, C. G., Cornel, M. C., Borry, P., Hastings, R. J., Fellmann, F., Hodgson, S. V., Howard, H. C., Cambon-Thomsen, A., Knoppers, B. M., Meijers-Heijboer, H., Scheffer, H., Tranebjaerg, L., Dondorp, W., and G. M. de Wert. 2013. "Whole-Genome Sequencing in Health Care: Recommendations of the European Society of Human Genetics." *European Journal of Human Genetics* 21:580–4.

Part II

Social Challenges

6 Risk and Solidarity within Individualized Medicine

Konrad Ott

Introduction

The following article emerged out of the GANIMED-research project at the Ernst-Moritz-Arndt University of Greifswald. Within the overall research project, there was a module dedicated to ethical and philosophical reflection on crucial concepts being used within the overall debate on "individualized medicine" (IM). The main results stemming from this module have recently been published elsewhere, e.g., in Fischer et al. (2015). I summarize some of the results in the first section, before turning to the main topic of this article in the remaining sections. My claim is, broadly speaking, that crucial features of IM in conjunction with health risk analysis support arguments in favor of a public health care system based on the principle of solidarity.

Individualized Medicine and its Critics

Reflection on IM has resulted in the conceptual insight that IM should be defined according to the interrelated concepts of stratification, prediction, and prevention. IM, at its very core (in its epistemic substance), is to be regarded as *stratifying, predictive*, and *preventive* approaches within medicine. At the moment, this approach is being intensively researched, especially on molecular levels. It remains an open question how IM will modify the healing practice of medicine in the future. This openness implies some leeway of how to design and implement IM and some responsibility for the chosen strategy for implementation. With regard to this responsibility, proponents of IM can profit and learn far more from its critics than from IM-enthusiasts and their nice bundles of promises.

Some criticisms of IM have been advanced by Marxists (Rajan 2009) and Foucauldians. One may learn from Marxism how investment strategies are pursued and new business models are to be launched by pharmaceutical companies (Ott and Fischer 2012). But the ethical foundations of Marxism in general are so shaky (to say the least), that one cannot ground a critique of IM in a detailed ethics of Marxism.[1] Some scholars have even denied that Marx himself grounded his critique of capitalism on moral reasoning (Wood

1986). Wood is a contemporary representative of that tradition of Marxist scholarship which strongly opposes moral reasoning as being ideological. To Wood, Marxists must only hold that a socialist economy will provide the majority of people with more non-moral goods (commodities) than a capitalistic economy. The contest between socialism and capitalism, then, would be reduced to a competition with respect to production and allocation of non-moral goods, including medical goods and services. After some decades of (clearly failed) socialist modes of production, we can judge the merits and shortcomings of both systems with some confidence. Contemporary Marxists may dislike the commercialization and privatization of medicine, but such criticism often remains on the rhetorical surface. The new IM-based medical business models are surely preferable when so-called "blockbuster drugs" are losing their patents. Even if in the take-off period some IM-products would find customers primarily among the well-educated, who enjoy higher salaries and private insurance, there might be a trickle-down effect over time. In the longer run (and only longer runs count in political economics), all or almost all citizens may profit from IM-products, as has often been the case within the history of medicine.

From a Foucauldian approach, IM can be conceived as an instance of "biopower." The concept of biopower sounds highly critical, and has moral overtones and political connotations, but within Foucault's writings, the concept of power is intrinsic to any kind of human interaction whatsoever. To perform agency is to impose power on other persons, be it in education, in politics, in economic exchange, or in medicine. The term biopower only denotes power within a realm of bio-politics (or bio-governance), which ranges from demography over medicine, epidemiology, social hygiene, to environmental protection. If so, the concept of (bio)power is a hybrid concept defined as an analytic-epistemic concept, but enriched by ethical suggestions and political connotations. Such enrichment constitutes the critical gesture of many figures in the Foucauldian camp. In line with Habermas (1985, ch. IX–X) and Honneth (1985, ch. 4–6), we (Ott and Fischer 2012) see severe ethical deficits in Foucault as well. At its best, a Foucauldian approach can make explicit the many tacit value-laden and crypto-normative premises within (a) medical discourses, and (b) practical interactions within medicine, but it does not provide a solid ground for an ethics of medicine. One also should take note that Foucault himself conceded modern biopower has effectively reduced mortality in modern Europe (Foucault 1986). If so, a critical philosophy of IM should neither be based on arguments against commercialization nor on biopower rhetoric.

In Ott and Fischer (2012, 2015), we argue that IM-philosophy can profit far more from Foucault's late interpretation of the ancient doctrine of *cura sui*. If we are embodied creatures, and if the body is prone to diseases, maladies, and disorders of all kinds, and if the difference between health and disease is intrinsically value-laden, one should take care of one's own health throughout one's life. *Prima facie* and *ceteris paribus*, we have sound reasons for staying healthy. Such a *cura sui* doctrine, if actualized for our present age, might turn IM into a "science of health" (Ott and Fischer 2015), which, on the part of individuals, relies on prudence and virtues. The IM-virtues, however, should not be based on sinfulness, discipline, rigor, shame, and guilt. Health care can and should instead be a life-affirming and even joyful activity. Such joyful *cura sui* will have to take the problem of health risks into account. The moral problem, however, remains how public health care systems should treat persons whose life styles are opposed to any kind of *cura sui*. Should they blame, shame, and penalize them or should they perceive them with generosity and forgiveness (see "Two Stylized Health Care Models" below)?

With respect to bioethical principles, I hold that the principle of informed consent (IC) remains largely intact within IM. In some respects, it remains not only intact, but gains in moral significance. It is, as I hold, valid *a fortiori*. Why? If IM deals with probabilities, dispositions, and tendencies which may or may not manifest over a lifetime, those seeking treatment should have the final say in how they cope with susceptibilities, disorders, risks, and the like. In other respects, however, the IC-principle should be modified according to the options and necessities of biobanking. If the charming ideal that all consent must be specific cannot be upheld within research based on biobanking, and if the practice of biobanking is highly promising for medical research, and if future hypotheses within research programs cannot be known in advance, the principle of IC might be "contractualized" with respect to data mining and data processing. Thus, the IC-principle should become more rigid in some respects and more liberal in others. Its core with respect to treatment decisions should remain intact. The same principle should also govern compliance with predictive checks and controls. Prevention should be based on voluntary, not on enforced, compliance. Since some risks of invasive diagnostics, and the possibility of "false positives" cannot be denied, one should be free to make use of the many predictive options available.

Within the medical practice of healing, the concept of commonly shared decision-making remains mandatory in a post-paternalistic paradigm of bioethics. Physicians should take the role of (empathic) supervisors and advisors, giving medical recommendations to presumptively prudent and autonomous clients. The interventionist impulse in medicine should be moderated, and "wait-and-see" strategies intrinsically belong to the portfolio of options within IM treatment. The post-paternalistic paradigm in medical ethics should be strengthened within IM. The reasons against paternalism are not refuted by IM.

To sum up, the value of health (axiology), a conjunction of life-affirming *cura sui* (eudaimonistic ethics), the principle of informed consent, and benevolent dialogical medical counselling within a post-paternalistic bioethical approach (normative bioethics) would seem to be appropriate for IM. IM can rest well on such solid ethical grounds.

There are impacts of IM not just on the level of bioethical principles, but on the more political and legal level of how to regulate and organize a public health care system (PHCS). Here, we face hopes and fears in relation to IM. While some hope for more individual responsibility on monetary terms (insurance), others fear an encroaching dissolution of the established health care system based in particular on solidarity with those individuals whose health status is rather poor; with handicapped individuals, persons with psychic disorders or chronic diseases, with members of marginalized groups, and the like. Here, PHCS is seen as an essential part of the welfare state, which should be protected, if not enlarged. Contemporary PHCS-debates are full of suggestive narratives, political rhetoric, shallow moralizing, economic calculations, and invalid interference. It is beyond the scope of this article to pick all the flaws in these debates, but it seems fair to say that (a) one cannot derive the normative principles of public health care systems from IM-definitions (Langanke and Fischer 2012), and (b) the concept of risk is crucial within debates on such principles. In the remainder of this article, I outline an argument based on some distinctions within the concept of health risks. By means of this argument, I hope to substantiate a principle of solidarity which is independent from political conflicts over welfare state policies within advanced (post)industrial society. Such independence is seen as advantageous.

Concepts Formation: Health, Risk, Solidarity

For a long time the concept of health has been a neglected concept of medicine, although it is supposed in theory and practice of medicine. Health is not a binary concept. Since Galen, philosophies of medicine have seen "full health" and "real disease" as polar opposites with many states in between them. "Full health" is less than the ideal state of complete physical and mental well-being which has been defined as "health" in the notorious WHO definition.[2] According to this (pseudo)ideal definition, almost all people are un-healthy at almost any time. Therefore, following Galen, I opt for a *scala sanitatis* between the two poles of real disease and full-blown health. Moving on such a scale, one also shifts between a more naturalistic and a more constructivist approach to health and maladies. In some sense, diseases are real states, while in other respects we stipulate what counts as disease and disorder. This dialectical intertwining of naturalistic and constructivist approaches seems inescapable. Moreover, I conceive human bodies as resilient organisms, which very often recover from a perturbation not just to a previous, but a new, state of health. Such recovery can bring about more or less health. Resilience comes and goes in degrees. If resilience vanishes, a person will only attain states close to the realm of sickness and maladies, if not disease. If human bodies have lost resilience, maladies become chronic, they intensify, and the organism dies. I am highly sympathetic to Galen's intuition that the "tonic fibres" of life, which mediate as vital forces between body and soul, must be exercised if the body is to remain healthy for longer periods.[3] From this perspective, resilient health can be maintained only through effort and struggle. Life, then, must be "forced" and "exercised." If so, a healthy life must to some extent involve risks.

Medical practices should, of course, cure diseases, but they should also strengthen organic resilience and the immune system,[4] as has long been advocated with the concept of salutogenesis (Antonowsky 1987). I assume more than just a grain of truth in such approaches. As Gadamer wisely argues in his *Die Verborgenheit der Gesundheit* (2010), staying in good health over decades, and swinging back again to states of health, should be a case for wonder, celebrating one's good fortune, and gratitude. This very sketchy concept of health is compatible with many nosologies, and it fits into the premises given in the former section. To my mind, IM stands in urgent need of a robust concept of health to counterbalance a preponderance to characterization in terms of diseases, maladies, disorders, syndromes, and the like. IM needs to discover the phenomena of resilience, strength, robustness, and recovery. I hope for a forthcoming medico-political coalition made up of IM-stakeholders and alternative concepts of health and healing. IM-theorists might also take a look at Nietzsche's inspiring notes on "great" health ("Große Gesundheit"). To Nietzsche, the term "great" indicates something comprehensive. "Great" health includes recovery from periods of sickness, robustness born from hardship, and courage based on a deep trust of one's body.

The term "risk" strictly speaking refers to courses of action which bring about upsides, such as joy, fun, preference satisfaction, but entail some likelihood of health impairments which might (or might not) occur. "Risk" is two-sided. In ordinary language, however, the likelihood of negative consequences (damage, harm, loss) is referred to as "risk." Following ordinary parlance, I define the strict notion of risk as "bet" (venture, "Wette") while conceiving the term "risk" with respect to the downside of a bet (Ott 1997). By definition, risk is always a combination of the *likelihood* of an event outcome, which counts as damage (loss, harm), with the *amount* ("meaning," gravity) of this damage. In ethical approaches toward risk, a highly important third feature of risk is its *distribution* over times and different agents. This feature should not be ignored in health risks since it is possible to impose health risks upon others, for instance via environmental pollution. Take the health risks of unskilled workers in tropical agriculture as a paradigm case. Purely individual risks, which affect no one besides the person who takes the risk, are exceptions. If one dies in solitary mountain hiking, friends and relatives will mourn. In our moral and legal practice, we have to balance the trans-personal character of risk taking with rule-based regulations of risks that include permissions, constraints, prohibitions, incentives, nudging, and recommendations of many kinds. Safety belts are mandatory, while helmets for cyclists are not; some vaccinations are mandatory, while others are not; some insurances are mandatory, while others are not. All in all, Western societies have reached a high level of safety in recent decades. Safety standards have increased. It is also likely that the many protests against hazardous technologies (nuclear power and deterrence, genetically modified organisms, pesticides, and the like) have also contributed to an increase in safety and security (Luhmann 1991). There are reasons to believe that from the historical perspective humans have never lived more safely than in contemporary Northern welfare states. An essential component of this safety is the potency of modern medicine. While overall safety has increased in our daily lives, we feel surrounded and threatened by megarisks such as climate change, biodiversity loss, global famine, and nuclear radiation. IM has to find its place in the overall culture of risk taking, risk aversion, and risk regulations with all its ambivalence and incoherence.

Risk taking is involved in exploring the world. Health impairments often emerge and occur in a time-delayed way, if at all. To distribute risks over time is often equal to *discounting* long-term risks. This might be acceptable with respect to individual risks, but a risk transfer to future persons (as in cases such as nuclear waste or climate change) seems repugnant. Whether discounting future events is irrational is contested within ethics and economics (see Hampicke and Ott 2003). Younger persons often discount risks which may manifest as damages later in life. Young adults cannot imagine themselves as being old although they "know" that everyone is aging. The rationale for discounting refers to the problem of whether a person will (not) regret her risky behavior in the distant future. We cannot know about our future regrets.[5] Individuals normally do not calculate the costs for their medical treatment decades in advance. Should they be forced to do so? To what degree should medicine accept that humans quite often behave myopically? Should the virtue of generosity entail some empathy with widespread myopia, and some forgiveness for non-compliance? It seems proper to add that increasing life expectancy has also increased the span of time within which a risk may manifest as health impairments. We would not assume that prolonging life expectancy is a major trigger of risks becoming manifest. Or should we? Can one say: "How lucky I have been to live long enough that such health impairments could manifest at all!" It is trivially true that the risk of mortality increases over time. An aging society wishing to reduce individual health risks is thus caught in a paradox. It makes IM a best aging contest. But paradoxes might be productive: They may increase the number of "good" years.

Some risks have clear probability, as in Russian roulette. If a risk has no clear probability, one should speak of *uncertainty*. Predictions within medicine are largely based on statistics, suggesting "objective risks." Quite often, however, both clients and physician face uncertainty. IM is clearly predictive and statistical. It deals with susceptibilities and probabilities. Thus, IM can provide clients with statistical information about the (un)likelihood of specific outcomes. The crucial distinction between *relative* and *absolute* risks is highly relevant at this point. All physicians should be familiar with this crucial distinction and all clients should be informed about it. That a relative increase of risk of, say, 30% is no absolute likeliness of 30% should be mandatory information. Without such information, clients confuse both kinds of risks and, by doing so, often overestimate risks. In IM, we are mostly dealing with both increase and decrease of relative risks.

Given both distinctions ("risk and uncertainty," "relative and absolute risk"), and given the statistical character of evidence-based medical research, health risks are often to be regarded as *relative uncertainties*. If so, any PHCS must take proper account of this crucial problem of relative uncertainties. IM must argue that uncertainties and relative risks should not affect the clear directedness of IM against myopia and non-compliance.

Granted this last point, I wish to suggest that we should distinguish between three types of health risks and uncertainties. Risks and uncertainties can be either (genetically) *given*, they can be voluntarily *taken*, or they can be *imposed* upon somebody by someone else (Ott and Fischer 2015, 157). There are hereditary legacies, there is one's own life style, which determines a specific risk profile being "voluntarily chosen," and there are externalities stemming from others. And there is a biography which mixes and accretes it all.

The three-fold distinction is close to common sense, and it is clearly non-reductive. It avoids the heroic assumption that all, or almost all, diseases have, ultimately, a genetic basis. IM can get along without such an assumption, although it tends to give some priority to the molecular research on hereditary risk factors being genetically given. One can, of course, try to identify genetic biomarkers via molecular biomedical research, but at the moment it often remains unclear what the voluminous data might indicate, and which interventions or even recommendations can or cannot be drawn from them. There is a risk of IM becoming lost in a jungle of fine-grained genetic biomarkers, which refer to relative uncertainties. Perhaps, embodied biomarkers, such as the power of the hands to grasp and to press things, are more significant than fine-grained molecular markers. Generally, we can firmly assume that genetically given dispositions are "folded" into social behavior in many complex ways (Buyx and Prainsack 2012a).

Intuitively, one would not make a person responsible for the hereditary risks given to her by genetic family lines. One's genome is a legacy one cannot change. No one is responsible for her own genome, but one might hold some specific responsibilities for her ways of coping with given genetic risks. Buyx and Prainsack (2012a, 81) see here a "widening of the range of scenarios and situations for which people are (at least partly) responsible." There is a tendency within IM to attribute responsibility even to genetically given relative uncertainties. One crucial question, then, is about moral accountability, or even legal liability to know about one's genome. The responsibility for acquiring genetic information is seen as a prerequisite of more specific accountabilities and liabilities. On this basis, one may even hold parents responsible for the genomes of their offspring. If so, it might be irresponsible to play the genetic lottery in the case of proliferation.[6] Some may believe that such responsibilities are "given" in roughly the same way as genetical risks are "given." Such a belief, however, would be a naturalistic fallacy. Despite these worrisome tendencies, the discourse on personal responsibility mainly refers to risks which have been taken voluntarily. Voluntariness presupposes that a person could have behaved otherwise.

This three-fold distinction is not perfectly clear. The case of *addiction* is an instance for the boundary zones between different kinds of risks. There is usually some initial responsibility for taking substances which entail the risk of becoming addicted to them, but there might be aspects of genetics and cultural imposition which also have to be accounted for. Perhaps, addictions to highly unhealthy substances are often more culturally imposed (by peer groups of young adults) or genetically triggered than freely chosen. Consider the risks of becoming a lonesome "workaholic." The borderline between risks taken and risks imposed is fuzzy, too. Many think we do what we do "willy-nilly" (*volens nolens*). How does the person with a well-paid job in a very noisy office, or a job with a commitment to pervasive high mobility and weekend-mailing, balance out the eustress and the distress along a professional career? Who is ultimately responsible for a burn out? Moreover, risks taken or risks imposed are volatile over a lifetime. One might have bravely quit unhealthy habits, but retain a slightly increased risk. Some risks are imposed upon us during childhood, and many risks are related to family life styles and diets, cultural patterns, and gender factors. Family habits incline toward risks imposed, as children tend to mimic their parents' behavior (Buyx and Prainsack 2012a). Thus, the adoption of healthy or unhealthy family habits falls somewhere in between "imposed" and "taken." How should we address the fact that females have, on average, a higher life expectancy than males although it is not intrinsically risky to be male? Male and female monks have equal life expectancies. Male sex is given, but the cultural gendering of being "male" may impose health risks upon males.[7]

Risks imposed can be seen as victimization. Risks imposed are relative to locations, rural and urban environments, working conditions, exposure to toxic substances in factories, on farmlands, and the like. Has the health risk of indoor pollution been taken by Nepalese housewives (preparing meals for the families), or has it been imposed upon them by a male-dominated caste society? Risks imposed upon people are clearly connected to social inequality, and to economic and political power relations. Working conditions in risky professions (such as mining) fall somewhere between risks taken and imposed, given the availability of alternative forms of employment.

The main point I wish to make is this: Every individual life comprises a highly individual (sic!) and literally concrete (from Latin *concrescere*) configuration of these three types of risks, which has been accumulated and processed over a lifetime. Such a concrete configuration is not just a "risk profile"; rather a risk profile is a focal perspective on specific aspects within the overall configuration. Let us dub this the "*g-t-i*"-problem: "*given*," "*taken*," "*imposed*," with all the fuzzy borderlines between them.

On the other hand, there are also many ways of counterbalancing health risks by health-promoting activities such as "healthy" diet, joyful play, recreation and leisure, (moderate) sports, (safe) sex, and the like, which intrinsically belong to *cura sui* practices. I would not even deny a certain top-down causality: Religious faith, attentiveness within meditation, and spiritual encounters with divine forces may have positive impacts on health. One may stay healthy if one simply trusts one's own body attentively. Even the borderline between "g-t-i" risky and health-promoting activities is quite fuzzy, as in the case of red wine, professional sports, sun-bathing, and the like. All in all, risky and health-promoting activities, with all their ambivalences, their contingencies, their closeness to good and bad luck, their fuzziness, their spreading over time constitute an individual health status. And, one may ask, may risks being taken even positively ("adaptively") contribute to the overall fitness and resilience of an embodied individual?

So we face both a configuration of "g-t-i" risks and a bundle of counterbalances, which are as volatile as the risk factors themselves. Both risks and counterbalances represent relative uncertainties. Given this, how can one determine the degree and proportion of health risks a person is justly responsible for? This question cannot be answered by facile reference to the popular examples of smoking, obesity, and unsafe sex. The real question is whether a PHCS should even attempt to calculate the proportion of individual responsibility for one's health? Would a PHCS that did be an improvement on the highly imperfect systems we have? In any case, a PHCS based on individual responsibility will be fair only if it takes account of both sides: "g-t-i"-risk configurations counterbalanced by health-promoting activities. This approach, however, would likely turn into an extremely complex model of incentives and rewards.

The concept of *solidarity* originally stems from the working-class movement. Solidarity has replaced the older revolutionary idea of "fraternity." In such genealogy, solidarity presupposes something which some groups of people have in common, such as commonly shared working conditions, salaries, daily lives, nourishment, exposure to unemployment, sickness, injuries, and the like. Factory workers shared a common fate. This "*sharing a common fate*" is the societal background of solidarity.[8] Due to this sharing, solidarity is always mutual on conceptual grounds (Zürcher 1998; Ott 2011). A wealthy tourist that gives some coins to an Indian beggar does not act out of solidarity since the two have nothing in common (except being human). The precise ethical status of solidarity in our socially dispersed and culturally plural society is no longer the clearly defined grouping with its origins in the working-class movement.

Solidarity is not a supreme moral principle. Neither is it a value, nor an individual virtue, nor a legal rule. It must be distinguished from altruism, compassion, generosity, and helpfulness. If I am helpful to a stranger in a low-cost situation, I do not feel motivated by solidarity. Solidarity has some binding moral constraints upon us. It is neither "nice" nor supererogatory, but mandatory to act in specific ways out of solidarity. If we act otherwise, we have earned the approbation of others. Such mandatory force connects solidarity to universal moral obligations, but in other respects solidarity remains particular. Universal solidarity with all humans in all respects is a somewhat strange ideal.

Without mutual trust, solidarity collapses. If we make divides between "them" and "us," solidarity vanishes (Buyx and Prainsack 2012a, 81). Precise calculations of costs and benefits may also undermine solidarity. In our case at hand, any (puritan) spirit of resentment against persons who seemingly enjoy unhealthy activities may also undermine the spirit of solidarity. If some people perceive some activities as not just unhealthy but vicious, solidarity is excluded.

Solidarity rests on some commonly shared fate, or on commonly shared vulnerability, proneness, or precariousness. At an interpersonal level, the notion of solidarity "comprises manifestations of the willingness to carry costs to assist others with whom a person recognizes sameness or similarity in at least one relevant aspect" (Buyx and Prainsack 2012a, 80). Such sameness or similarity constitutes *fellowship*. Commonly shared solidarity is constituted by some recognition of fellowship. Recognition of fellowship implies the attitude of readiness and willingness to give assistance to any fellow in specific cases of need and emergency. If, on the other hand, I am the one who is deleteriously affected or stands in need, I must be able to trust and rely on the readiness and willingness of my fellows to assist me. Recognition of commonly shared proneness seems crucial for solidarity. A paradigm case might be the commonly shared risk of finding oneself in an emergency situation in which one can no longer help oneself. One cannot escape without assistance. Such a situation may befall any one of us, but we do not know in advance to whom it may actually happen. Such recognition of commonly shared vulnerability seems to be a necessary condition for actual solidarity.[9] The decisive criterion of relevant similarity might be, of course, very general or very specific.

In our case at hand, commonly shared vulnerability is the general exposure to disease, malady, injuries, and the like. We know about such vulnerability through life experience if we have come close to death or suffered severe injuries. Health, however robust, remains fragile. Perhaps, every drop of blood reminds us of such organic vulnerability. All other embodied individuals are, in principle, as vulnerable as oneself. To emphasize this common fate, however, does not nullify the many differences. We all are prone to maladies, but proneness to lung cancer is far higher for smokers. What about the proneness to Morbus Parkinson's in an aging society: Should intellectuals feel safer than other people? The case for solidarity lacks an argument as to why the general common human proneness to maladies outweighs the many differences in likelihood. In my view, the g-t-i approach to health risks as given in the previous section provides such argument.

In summary, solidarity is about mutual expectations of assistance in the face of common vulnerability. Contractarian calculations (*do ut des*), however sophisticated, miss the ethical grounding of solidarity. If this is generally true, it might be also true in the field of PHCS.

Two Stylized Health Care Models

The problem of PHCS is about *whether*, and if so, *how* to organize solidarity among individuals who are united under the commonly shared exposure to diseases, but remain anonymous strangers to each other within insurance systems based on monetary contributions. I will deal with the basic question of whether first, and then turn to the question of how. The two models being laid out in this section are ideally stylized. No real public health care system will resemble either of them exactly. Real systems may, however, differ according to their organizing principles.

Calculation

There are many complaints about cost increases in health systems. Meanwhile, Western societies spend more than 10% of GDP on health care services. In the U.S., it is roughly 15%, with uneven access to such services, but a high-end health care for the wealthy. In aging societies, the proportion of GDP is unlikely to decline, but there will be an upper ceiling because there would be heavy trade-offs if the health care system were to consume, say, 25% of GDP. Individual responsibility for one's own health care is often extended into culpability for general cost increases within PHCS. Complaints over cost increases and criticism of irresponsible behavior are often conflated. Thereby, irresponsible behavior is conceived as a major driver of cost increase in PHCS. It is argued that such behavior may make insurance bills higher than they would be if all or almost all clients and patients behaved (more) responsibly (according to specific standards and criteria to be determined).

Hopes that medical innovation may decrease costs have mostly been in vain. There is, however, new hope with respect to IM, whereby it is argued that prevention might be less costly than treatment. Treatments can be avoided through prevention. If resources were invested in IM-based prediction and prevention, a substantial reduction in costs for treatment and cure might be achieved, if and only if most individuals comply with a stratified, predictive, and preventive system. In any case, general compliance is crucial for prospects of cost decreases. This scheme of calculation is, of course, very rough. It abstracts away many other drivers of costs within PHCS (such as patents, salaries, administration, etc.). It seems fair to say that individual behavior is only one driver of costs among many. If IM is a macro-innovation in medicine, it remains an open question how cost structures will be affected on the macro-, meso-, and micro-levels of PHCS (Flessa and Marschall 2015). Drivers of costs within PHCS are to be analyzed with scrutiny and caution.

The point I wish to make is the following: The popular narratives of irresponsible behavior might pave the way to a PHCS within which health risks are assessed on an individual basis according to stratification, prediction, preventive strategies, and compliance.[10] IM serves as an epistemic justification for such a PHCS. Let us call this approach: *calculation & privatization* (C&P). The C&P approach will be attractive for new business models, as well as a never-ending research program, and, therefore, might find influential proponents. One crucial feature of such a C&P-system might be this: *The epistemic stratification of IM is to be prolonged into PHCS via economic stratifications of insurance policies*. The C&P-based PHCS only seems to execute the epistemic insights stemming from advanced IM-research. In such a PHCS, monetary incentives and disincentives become the usual means of enhancing compliance, and insurance bills are calculated precisely with respect to individual risk profiles. This trend "toward more risk stratification in private insurance" (Buyx and Prainsack 2012a, 81) is strong within liberal variants of capitalism with a cultural background in puritanism, as in the U.S. As such, costs of medical services would be individualized and privatized according to appropriate standards of behavior and risks.

In such a system, individuals are conceived of as isolated subjects of health care. On average, they wish to optimize their benefits as members of PHCS. They wish to stay in good health as long as possible, do not wish to pay high insurance bills, and are willing to comply. In a nutshell, most ordinary members of the PHCS wish to live healthily and pay less for their health insurance. They do not wish to pay for the unhealthy, vicious, or even "sinful" activities of persons they disapprove of. They do not wish to pay for the irresponsible behavior of anonymous others, and dislike freeloading and cheating.

Thus, C&P requires honesty and compliance on the part of the client. Cheating would have to be penalized. This would require control of behavior, and detection of cheating and non-compliance in a scientifically-based way. In such a C&P-system, there would be a multitude of contractarian options. On the one hand, the system would provide liberty of choice: "It's up to you. It is your personal choice. It's your precious life." On the other hand, it would be asymmetrical with respect to information and economic power between insurance companies and individual customers and clients. Governance schemes would have to be established to balance the interests of customers and insurance agencies.

A fundamental problem remains with regard to how to calculate one's overall "g-t-i" risk configuration of the specific relative risks a client is justly responsible for. By which standards, criteria, and indicators can someone be held responsible for her disease? How should responsibility be related to insurance contracts? Should *cura-sui* behavior be reflected in monetary terms? It might be unfair within a C&P-system to penalize irresponsible behavior but not to reward efforts to stay healthy. Such calculations would not be purely scientific, but would be full of arbitrary assumptions, cultural biases, vested interests, hidden agendas, and high transaction and control costs. One crucial point is the transformation of uncertainties into risks, since only risks can be monetized. Under such a C&P approach, we are faced with several epistemological and ethical problems:

- Causality and dispositions with respect to g-t-i configurations
- Compliance and surveillance, control, early detection
- Distributive justice and access to medical services
- Incentives and rewards, "nudging"
- Counselling and advisory strategies
- Respect for the rights of dissenting persons
- Appropriate governance schemes
- Transformation of uncertainty into risk

Given all the scientific uncertainties of prediction and probabilities, such a C&P-approach is, at best, a thorny enterprise. It is debatable whether the C&P-model is even, on closer inspection, attractive to the pharmaceutical industries and to insurance agencies. In my own view, a C&P-system of this kind would be abhorrent. This abhorrence is not mere moral indignation. It is more complex.

From a cultural perspective, C&P is based on the puritan suggestion that health risks such as over-eating, excessive drinking, smoking, unsafe sex, and taking drugs are always close to "sinful" activities. To puritans, there are no moral reasons to feel solidarity with sinners. And why should decent persons who reduce their sinful activities pay higher insurance premiums for the sinful activities of others? C&P fits well into a puritan culture in which sins ought to be punished. Thus, there are reasons why such a C&P-approach finds much support in variants of liberal capitalism and a cultural background in puritanism.

From the perspective of moral phenomenology, it is disgust at being forced or nudged into conformity with a system that is far less reasonable than it presumes to be. From the perspective of critical theory, such a C&P-approach might even turn out to become an instance of an eclipse of reason and the dialectics of enlightenment (Horkheimer and Adorno 1947), resulting in a Kafka-esque dystopia within biomedicine. Every health risk must be calculated and monetized precisely, but the system as such becomes absurd. The social movement of a quantified self, whose members permanently check their health parameters, may indicate the "uto/dystopian" direction of the C&P-approach. Ultimately, perhaps, the great IM buzzword, namely "individuality," might collapse into well-controlled physiological parameters, and a precisely calculated insurance bill. I leave it for further debate whether IM scientists would really feel comfortable with and within such a PHCS.

Solidarity

In the first instance, we can and should perceive others as fellows united by commonly shared exposure and vulnerability. Everyone has a specific profile and burden of g-t-i risks. The many specific features are less significant even if we recognize that each risk profile is specific. Thus, the most common denominator of proneness to maladies remains. Under such broad criteria of similarity and fellowship, the tempting idea of extending epistemic stratification to economic stratification of insurance policies should be discarded. Although IM is to be defined via concepts of epistemic-medical stratification, a PHCS should resist the temptation to stratify clients and their insurances (Buyx and Prainsack 2012a).

Most continental PHCS provide good or decent provision of medical services even to the least advantaged persons. Most people just take it for granted that those who live by welfare-state transfers are insured without actual payments,[11] but we should not overlook the actual solidarity implicit in such institutions. A solidarity-based approach (SB approach) conforms to our continental traditions of welfare states and, perhaps, even to Christian traditions of giving special attention to *personae miserae*. It is a simple truth that PHCSs are not primarily designed for the young and healthy. Such an SB approach also conforms to the egalitarian intuitions many Europeans feel with respect to access to health-care services. Monetary income should not be decisive in matters of health and sickness, life and death. Behind a Rawlsian veil of ignorance, prudent and risk-averse persons would adopt a health-care principle analogous to the principle of difference: Knowing one's own vulnerability but not knowing one's societal position provides a strong rationale for a minimax-strategy. If the veil is lifted, one may find oneself handicapped or struck by a chronic disease. Thus, Rawlsians should clearly favor an SB system. Individual *cura sui* and solidarity are clearly compatible with each other, and with the acknowledgement of any one's g-t-i configuration.

SB approaches can also be supported with respect to some crucial virtues within an ethics of medicine. Virtues such as generosity, mercifulness, and forgiveness can be regarded as virtues accruing to individual physicians, but they might be extended to the attitudes of fellows within an SB system. Solidarity entails an intrinsic relation to such virtues. Without such virtues, however, solidarity might collapse into a contractarian concept of prudent persons who decide to found a club whose members share access to "club goods."[12] I wish to focus on generosity and forgiveness.

Solidarity and generosity have something in common: They both refuse to calculate sharply. The attitude of solidarity does not even expect that individual payments in PHCS budgets and the "take" will balance over time. Some people will, indeed, in the longer run pay more and take less than others, as the permanently employed pay far more into unemployment insurance than they take out. Social security systems are stabilized by persons who need them less than others. But within an SB approach this is not a reason to complain. On the contrary: Those who pay more than they take out of the budget *ceteris paribus* should feel happy of being healthy and employed most of the time. Healthy persons should rather pity their fellows standing in need of repeated medical treatment in a non-judging, empathic way.

Under a solidarity approach, we should recognize each other with compassion and forgiveness. We should be aware of the temptations and longings which induce health risks of different kinds. We should recognize each other as non-perfect beings in a highly non-ideal world full of aspirations, longings, projects, and visions. It is not only unsafe sex, risky sports, legal or illegal drugs, it is also professional competition, careers, and neglect of symptoms that lead to ill-health. We know about the weakness of the will (*akrasia*), the slippery slope toward addiction, and the role of good or bad luck in health affairs. We take a deep interest in our health but we are not "health heroes" in our daily lives. We have to acknowledge each other as individual configurations of g-t-i uncertainties with forgiveness. Forgiveness is the contrary of blaming and shaming. The conceptual relation between solidarity and forgiveness is not merely additional, it might be intrinsic. Without forgiveness, solidarity collapses back into contractarian prudence of membership within a club. In any case, an SB model does not rest on the assumption that risks taken are analogous to sinful activities which must be punished by extra payments. In SB approaches, solidarity is solidarity among "sinners" who forgive each other the risks they have taken. But should we forgive, one may ask, some, most, or all of such risks? May forgiveness result in morally hazardous behavior, if all "sins" are to be forgiven? Forgiveness stretches the limits of our solidarity. But how far does solidarity reach, if solidarity always has to face limits somewhere?

A solidarity-based approach has to face serious problems. I will mention just two of them. Firstly, solidarity is scarce, fuzzy, and it is never safe against misuse by free-riding. One has to protect such an SB system against the misuse of freeriding, morally hazardous behavior, and rent-seeking. "Being sick" is, at least to some people, an all-too-human strategy to cope with the hardship of life in societies based on labor, mobility, and competition. How far does generosity reach if persons adopt such strategies to slip out of the system of economic cooperation? Secondly, a solidarity-based approach under such a common denominator is, *in principle*, without limit. It also includes, in principle, poor people in other countries, and might not even be constrained to humans since animals are prone to maladies, too. People in absolute poverty suffer far more from disease burdens than Western people.[13] This article may be accused of restricting solidarity to "solidarity among the privileged." Since global solidarity in health care is beyond my horizon, and because I am loathe to present idealist wishful thinking, I will, with some reluctance, remain silent on global solidarity.

Conclusion

Progress in medical research paves the way toward a new paradigm of IM medicine based on stratification, prediction, and prevention. IM will swim and dive in the vast ocean of susceptibilities. Scientific careers will be devoted to molecular biomarkers. The times seem to change rapidly in the field of bio-medicine. As some believe, rapid changes in science may imply profound change in morals and even ethics. This might, however, turn out to be a *non sequitur*.

Health will remain a deeply appreciated, valuable, and joyful state for human beings. The ancient concept of *cura sui* may well have a future within both an aging society and the emerging IM paradigm. IM should coalesce with prudence in dietary health care and with "wellness activities." IM should not imply a return to a paternalistic approach with respect to the relationship between physicians and clients. The principle of informed consent remains valid. Moreover, a nuanced approach to health risks based on the g-t-i distinction strengthens the principle of solidarity and, perhaps, even supports virtues such as forgiveness and generosity. These attitudes accept the weaknesses of the human condition to some degree. Such an ethics of IM is by no means revolutionary. It connects personal liberty, individual prudence, solidarity based on fellowship, and, ultimately, an outlook on new visions of human health.

Notes

1 For an overview of the debate on ethics and Marxism, see contributions in Angehrn and Lohmann (1986).
2 For one of many critical comments, see Wehkamp (2012).
3 This point has been outlined by Prof. Kuriyama (Harvard) in a lecture on the "Riddle of Presence" delivered at Kiel University in May 2015.
4 See for the concept of "psychoneuroimmunology" Hyland (2011), especially chapter 2. At the end of his inspiring scientific book, Hyland notes: "To be healthy one needs to enjoy life" (2011, 303). Perhaps, enjoying can encompass biophilic attitudes and, ultimately, even some reverence for life as such.
5 Continuous health care supposes some unity of a person over time. The predictive force of IM would be shattered if persons hold disuniting concepts of personhood. Why should I care for a cancer in the remote future, if I do not believe in a self that continues over time? For the sake of argument, I discard the option of a disuniting concept of personhood.
6 Here, a slippery slope toward "liberal" eugenics looms. But, as with any slippery slope, one may ask what is morally wrong with prenatal eugenics if a substantial reduction of genetic risks can be achieved? One may stipulate a "right not to know" at this point, irrespective of how many clients wish to make use of such a right with clinical practices. I will not address this topic.

7 Why do many of us believe that it is a moral scandal if the poor live shorter lives than the rich while none of the many gender-experts makes a moral scandal out of the fact that gendering "male" implies shorter lives for males?
8 This section relies on Buyx and Prainsack (2012a, b).
9 But is such recognition also in itself sufficient? Why should we not calculate which vulnerability may apply to us with some degree of likelihood? Does it make sense to state that one feels solidarity (not just compassion) with persons who are extremely poor? I leave this point for further investigation.
10 Such narrative should make one skeptical about a "narrativistic turn" in ethics.
11 In Germany, such persons are "*beitragsfrei mitversichert*".
12 Such prudence, in reality, has been a root cause for a decline in common solidarity, since professional groups (physicians, lawyers) have founded such special clubs. An underlying moral problem is this: Why should such special clubs be repugnant if it is permissible to establish a club based on citizenship? This question I feel unable to answer.
13 Membership in a PHCS, however, is reserved to co-citizens even if all humans are prone to maladies. Real PHCSs are "clubs" and there is much malady and misery outside the "clubs". We should be aware of the privileges we enjoy in Western PHCSs. Under such constraints and privileges, we can and should give free access to health-care services to poor co-citizens and to the refugees within our borders, and should donate resources for establishing health-care facilities in least developed countries. However, these topics are beyond the scope of this article.

References

Angehrn, E., and G. Lohmann, eds. 1986. *Ethik und Marx: Moralkritik und normative Grundlagen der Marxschen Theorie*. Königstein: Athenäum.

Antonowski, A. 1987. *Unravelling the Mystery of Health*. San Francisco: Jossey-Bass.

Buyx, A., and B. Prainsack. 2012a. "Lifestyle-Related Diseases and Individual Responsibility through the Prism of Solidarity." *Clinical Ethics* 7:79–85.

Buyx, A., and B. Prainsack. 2012b. "Solidarity in Contemporary Bioethics—towards a New Approach." *Bioethics* 20(7):343–50.

Fischer, T., Langanke, M., Marschall, P., and S. Michl, eds. 2015. *Individualized Medicine*. Heidelberg: Springer.

Flessa, S., and P. Marschall. 2015. "Individualized Medicine: From Potential to Macro-Innovation." In *Individualized Medicine*, edited by T. Fischer, M. Langanke, P. Marschall, and S. Michl, 253–91. Heidelberg: Springer.

Foucault, M. 1986. *The Care of the Self: The History of Sexuality*, Vol. 3. New York: Random House.

Gadamer, H.-G. 2010. *Die Verborgenheit der Gesundheit*. Frankfurt/M.: Suhrkamp.

Habermas, J. 1985. *Der philosophische Diskurs der Moderne*. Frankfurt/M.: Suhrkamp.

Hampicke, U., and K. Ott, eds. 2003. Reflections on Discounting. *International Journal for Sustainable Development*, Special Issue 6(1) 1–149.

Honneth, A. 1985. *Kritik der Macht*. Frankfurt/M.: Suhrkamp.

Horkheimer, M., and T.W. Adorno. 1947. *Dialektik der Aufklärung*. Amsterdam: Querido.

Hyland, M.E. 2011. *The Origins of Health and Disease*. Cambridge: University Press.

Langanke, M., and T. Fischer. 2012. “Gesundheitsmanagement liegt mir im Blut: Individualisierte Medizin und gesundheitliche Eigenverantwortung.” In *Konzepte des Humanen*, edited by M. Gadebusch Bondio and H. Siebenpfeiffer, 139–72. Freiburg: Alber.

Luhmann, N. 1991. *Soziologie des Risikos*. Berlin: DeGruyter.

Ott, K. 1997. “Ethik und Wahrscheinlichkeit.” *Nova Acta Leopoldina* 77(304): 111–33.

Ott, K. 2011. “Solidarität in Europa und der Fall Griechenland.” *Vorgänge* 196(12): 34–47.

Ott, K., and T. Fischer. 2012. “Can Objections to Individualized Medicine Be Justified?” In *Individualized Medicine between Hype and Hope*, edited by P. Dabrock, M. Braun, and J. Ried, 173–200. Münster: LIT.

Ott, K., and T. Fischer. 2015. “On a Philosophy of Individualized Medicine: Conceptual and Ethical Questions.” In *Individualized Medicine*, edited by T. Fischer, M. Langanke, P. Marschall, and S. Michl, 115–63. Heidelberg: Springer.

Rajan, K.S. 2009. *Biokapitalismus: Werte im Postgenomischen Zeitalter*. Frankfurt/M.: Suhrkamp.

Wehkamp, K. 2012. “Gesundheit als Potenzial. Warum der WHO-Gesundheitsbegriff verändert werden sollte!” In *Konzepte des Humanen*, edited by M. Gadebusch Bondio and H. Siebenpfeiffer, 103–16. Freiburg: Alber.

Wood, A.W. 1986. “Marx‘ Immoralismus.” In *Ethik und Marx*, edited by E. Angehrn and G. Lohmann, 19–35. Königstein: Athenäum.

Zürcher, M.D. 1998. *Solidarität, Anerkennung und Gemeinschaft*. Bern: Francke.

7 Anticipatory Medicalization

Predisposition, Prediction, and the Expansion of Medicalized Conditions

Peter Conrad and Miranda Waggoner

For nearly half a century sociologists and others have examined the increasing medicalization of society (Zola 1972; Conrad 2007; Conrad 2013) with the goal of understanding the origins, expansion, and consequences of medicalization. Some critics see most medicalization in a more negative frame as overmedicalization, and as a societal problem (e.g., Illich 1975). More balanced commentators, however, examine medicalization as they would any other major social trend (e.g., urbanization, industrialization, secularization, etc.), and evaluate the social consequences of these changes. One point most analysts would agree upon is that in the past fifty years there has been a significant increase in the medicalization of human conditions. As we discuss below, there are a number of ways in which medical diagnoses and treatments of problems have expanded. In this paper, we will examine what we call "anticipatory medicalization," which medicalizes a condition *before* a problem or condition has manifested. But prior to venturing into this new territory, we will describe the characteristics of medicalization and examine several ways in which medicalization has expanded in the past fifty years.

What Constitutes Medicalization?[1]

While there may be numerous definitions of medicalization, in its simplest form it means "to make medical." A more formal definition sees medicalization as the process by which previously nonmedical problems become defined and treated as medical problems, usually as diseases or disorders. The emphasis in this perspective is on "process" and "definition."

Medicalization is distinguished by several characteristics. (1) The *definitional* issue is central to medicalization; that is, how a problem is defined is key to what is done about it. While a medicalized condition usually is defined by physicians or medical personnel, this need not be the case. (2) Medicalization can be seen as representing a range of definitions, with some problems totally medicalized, and others just barely medicalized. This suggests *degrees* of medicalization: for example, some problems are virtually completely medicalized (e.g., schizophrenia, epilepsy), while others, such as Attention Deficit Hyperactivity Disorder (ADHD) are mostly medicalized, or, as in the examples of internet addiction and sexual addiction, are barely medicalized, with still others contested or somewhere in between (e.g., obesity, opiate addiction). In short, medicalization is more of a continuum than a binary either/or distinction. (3) Medicalized categories are *elastic*, and can expand or contract. Two examples of this are ADHD and Posttraumatic Stress Disorder (PTSD), which will be discussed later in this paper. Both have expanded their diagnostic reach since they were introduced. On the other hand, hysteria, which was a common medical category in the late 19th century, has shrunk and almost disappeared as a diagnosis. (4) Physician involvement in medicalization is variable. With regard to the medicalization of alcoholism, for example, physicians were minimally involved; rather, the disease ideology of Alcoholics Anonymous was central to this process. Consumers can promote medicalization as well, requesting or promoting medical definitions and treatments for problems. With many other examples, however, some direct medical involvement is key. (5) Medicalization is *bi-directional*; that is, there can be medicalization as well as demedicalization. The most common example of demedicalization has been homosexuality, which due to the 1974 vote by the American Psychiatric Association, and subsequent changes in attitude, became "officially" demedicalized. There is no question, however, that there has been much more medicalization than demedicalization in the past four decades.

Sometimes we see the medicalization of "old" problems, other times medicalization of newly discovered problems or conditions. There is no question that some conditions have been long defined as problems, and in our current age have become medical problems. For example, difficulty in becoming pregnant has been long seen as a human problem, and those affected sought help of some kind. One need only view the "fertility objects" that are readily found in any anthropological museum to recognize the ubiquity of the problem. However, since the invention of in vitro fertilization (IVF) in 1978, infertility has been increasingly medicalized in Western society, and IVF treatments are now common practice. Additionally, such examples might include the medicalization of obesity, aging, and sleep. On the other hand, there are conditions that become medicalized *largely because there is a treatment available*, for example, idiopathic shortness and human growth hormone (Conrad and Potter 2004), shyness as social anxiety disorder with Paxil as the treatment (Lane 2007), and, to a degree, school behavior problems becoming defined as ADHD with the advent of Ritalin in the 1960s (Conrad 1975). Other examples could include menopause and hormone replacement therapy, Viagra and increased diagnosis and treatment of occasional erectile dysfunction (Loe 2004), and the expansion of normal sadness into diagnosed depression in the Prozac generation (Horwitz and Wakefield 2007). In these cases, it is the availability of a medical "treatment" that underlines the emergence of a medical definition or diagnosis for a formerly "normal" condition or behavior.

The Expansion of Medicalization

While nearly all writers on medicalization agree that medicalization of human conditions and problems is increasing (e.g., Clarke et al. 2003; Conrad 2007; Bell and Figert 2015), most point to the increase in the number of diagnoses of conditions that have become medicalized. Some examples in the past few decades include obesity, Internet addiction, gender dysphoria, infertility, Premenstrual Syndrome, shyness (social anxiety disorder), PTSD, erectile dysfunction, Fetal Alcohol Syndrome, Alzheimer's disease, obsessive compulsive disorder, and many others. Many of these conditions were previously seen as problems in society, but more recently have become defined as medical problems with diagnoses and medical treatments. There is little doubt that new diagnoses for existing conditions represent a major way in which the medicalization of society is growing. For example, the fifth edition of the *Diagnostic and Statistical Manual of Mental Disorders* (DSM-5) removed the term Asperger's Syndrome, and expanded definitions of depression to include bereavement and grief (Bandini 2015). But there are other modes of increasing medicalization.

In addition to creating new diagnoses, there are also various possible forms of *diagnostic expansion*—that is, broadening the criteria for a diagnostic category. ADHD provides an interesting example. When ADHD was first described as "minimal brain dysfunction" or "hyperactive syndrome," the emphasis was on hyperactive and impulsive behavior. In the 1980s, when the diagnosis became ADHD, the diagnostic focus changed to "inattentive behavior," and then not only were "hyperactive" children diagnosed, but "spacey," inattentive children as well. This widened the threshold of diagnosis. When hyperactivity and ADHD were first diagnosed, it was believed that the disorder was mostly limited to 6–12 year olds, and that many children outgrew the disorder. But over the years we began to see diagnoses expand by age to include adolescent ADHD, adult ADHD, and now toddler (as young as 4) ADHD, so that ADHD has become a lifespan disorder with virtually no age limits (Conrad and Potter 2000). In the 1970s, it was thought that ADHD affected about 3% of the population, but now, with the expanded diagnosis, the latest Centers for Disease Control and Prevention (CDC) estimate is 11% of the population (Schwarz and Cohen 2013).

Another example of a type of diagnostic expansion can be seen with PTSD. PTSD emerged as a diagnosis for Việt Nam war veterans who returned home with symptoms of anxiety, difficulty sleeping, fear-driven or unpredictable behaviors, social withdrawal, and especially "flashbacks." Within a decade the diagnosis was applied to numerous other "post traumatic" situations: survivors of child abuse, sexual abuse, rape, and violence; survivors of earthquakes or other severe natural disasters; and even those who had witnessed violence, abuse, or natural disasters. What started out as a very specific psychiatric diagnosis grew to be a common diagnosis for a whole range of stressful life experiences.

The actual number of individuals designated with a disorder can increase when a diagnosis is used with a new population. One way this can occur is when a diagnosis migrates from one country to other countries. For example, until the 1990s ADHD was predominantly diagnosed in the U.S., Canada, and Australia. When in the 1980s U.S. physicians were diagnosing 5% of school children, in the U.K., fewer than 1% were diagnosed (Hinshaw 1994). But in the last two decades the ADHD diagnosis has migrated to dozens of other countries, significantly increasing the numbers of children diagnosed and treated with the disorder, and contributing to the global medicalization of children's difficulties that are seen as ADHD (Conrad and Bergey 2014).

Thus we can identify several ways in which expanding diagnoses can increase medicalization. In the remainder of this chapter, we will introduce the concept of "anticipatory medicalization" and a particular example (usually referred to as preconception care) as another mode of increasing medicalization.

Anticipatory Medicalization

In the trend toward the expansion of medicalized conditions, we also highlight a shift toward focusing on the future, i.e., toward anticipating the medicalization of conditions. Anticipatory medicalization revolves around the expectation of a medical diagnosis or medical outcome; it depends not on a present condition, but rather on putative potential problems. Health risks may or may not be visible or detectable now, but in the framework of anticipatory medicalization, the clinical concern is about future risks. We conceptualize anticipatory medicalization as defining and/or treating a putative potential problem with medical interventions because it may pose a risk in the future.

To illustrate anticipatory medicalization, we use the example of the rise of "preconception care" in medicine, or the idea of caring for non-pregnant women with the purpose of alleviating any risks to future pregnancies. Throughout the course of the 20th century, medical knowledge increasingly suggested that clinical care and healthy behavior during the nine months of pregnancy defined the cornerstone to achieving best birth outcomes and reducing infant deaths (Barker 1998). Yet, as more women accessed prenatal care services, improvement in birth outcomes stalled. Despite this reality, health initiatives in the last two decades of the 20th century still strove to enhance women's awareness of the importance of prenatal care through prenatal education (Armstrong 2000). In response, some maternal and child health (MCH) experts initiated a new focus on the period *prior to* conception. Among many obstetricians and MCH specialists, the idea emerged that women should be at optimal health prior to conception in order to assure the best pregnancy and infant outcomes in the future (Johnson et al. 2006).

This shift toward a medicalized definition of the pre-pregnancy period occurred in the context of the medicalization of reproduction. Reproductive events have long served as especially illustrative examples of the medicalization process, particularly topics related to the surveillance of pregnant and birthing women's bodies and behaviors (e.g., Wertz and Wertz 1977; Rothman 1993; Armstrong 2003). Feminist studies on these topics highlight the expansion of medical surveillance over women's reproductive bodies throughout the course of the 20th century. Pregnancy, for instance, was deliberately re-conceptualized by health professionals as a biomedical and pathological event, rather than a social and physiological one (Barker 2003)—and with specific knock-on effects. At the beginning of the 20th century, women in the U.S. did not seek prenatal care and gave birth at home. In contemporary U.S. society, almost all women receive clinical care throughout their pregnancies and travel to a hospital to give birth. The medicalization of pregnancy followed the rise in medical interpretations of pregnancies, such as in the dissemination of health and behavioral advice to pregnant women (Barker 1998), and in medical interventions during pregnancy and childbirth, such as with the advent of prenatal screening techniques and the overuse of caesarean section. Today, the medical surveillance of the lives of virtually all women of childbearing age—as is the case with preconception care—signals a marked expansion of medicalization.

The impetus for a new definition of reproductive risk and pregnancy care in the 21st century was forged with the emergence of the notion that clinical care prior to pregnancy might result in better pregnancy outcomes. The 21st-century model for maternal and child health care—in the United States, but increasingly in European countries as well—is to routinely screen women of reproductive age for a variety of health conditions, such as diabetes or obesity, in order to assess possible risk factors pertaining to the health of future pregnancies. Pregnancy health risk is thus defined as occurring *before* pregnancy, and the overriding goal is to reduce the (future) risk of adverse pregnancy and birth outcomes, such as preterm birth, low birthweight, and infant mortality.

Preconception care interventions specifically focus on reducing any risk factors that may in some way influence the health of a future pregnancy. For instance, women are urged to undergo screening in order to assess the presence of any genetic predispositions, or of any sexually transmitted infections. Clinicians and public health experts advise women to take folic acid supplements, and women are generally counseled to control any chronic diseases (such as diabetes), and to stop smoking or drinking. This surveillance approach has been operationalized in clinical medicine as an "every woman, every time" outlook, in which physicians are urged to evaluate potential reproductive risks at every clinical encounter (Atrash et al. 2008). In practice, this treatment model defines all women as though they are potentially pregnant, and thereby anticipates the medicalization of pregnancy and reproduction (Waggoner 2013). Although it is not always clear from the medical literature that this form of prevention is an effective one, it is apparent that this new treatment framework represents a form of expanded medicalization.

To be precise, we argue that the example of preconception care characterizes anticipatory medicalization. Risks may or may not be actually present in any presenting female patient of childbearing age; rather, the clinical concern is about future risk her body may pose to a potential pregnancy. This focus creates anxieties about the future in addition to engendering a new orientation to the patient, one that revolves around putative potential problems. A non-pregnant woman of reproductive age who is not in optimal health may not pose any serious risks to her future fetus. However, by defining her non-pregnant body as potentially risky to her future fetus, the medicalization of reproduction is anticipated. This is a way in which we see the continued expansion of medicalization, toward locating any potential conditions in the future.

Anticipatory medicalization is related to previous conceptions in medical sociology. Scholars have paid attention to ways in which medicalization has expanded through increased focus on prediction, predisposition, and proto-disease. Prognosis—or predicting the outcome of a condition—constitutes a main task for physicians, and it is often quite variable in its delivery and accuracy (Christakis 1999). Medical sociologists and historians of medicine have focused in recent years on the trend in medicine toward managing the uncertainty and fear associated with the liminal space between normalcy and pathology, treating healthy populations as if they were primed for illness, before a condition is even present and in need of prognosis (Armstrong 1995; Aronowitz 2009; Rosenberg 2009). In some ways, this medical trend may represent an overextension of prevention. As individuals are increasingly designated to have a predisposition for a certain disease or outcome, such as pre-diabetes or pre-hypertension, we are witnessing the rise of the pre-symptomatic individual (Clarke et al. 2003; Konrad 2003; Conrad 2005). Adams et al. (2009) write that "anticipation is rapidly reconfiguring technoscientific and biomedical practices as a totalizing orientation" and that "anticipation pervades the ways we think about, feel and address our contemporary problems" (2009, 248).

Other studies have presented examples that unearth the anticipatory nature of medicalization in contemporary society. In a study of men with elevated prostate-specific antigen (PSA) levels, Gillespie (2012) reveals how the label of a pre-disease state produces anxiety among patients. Charles Rosenberg uses other pre-disease risk states, elevated cholesterol, or hypertension, to refer to "proto-disease states" (2007). The advent of chemoprevention, as another example, includes administering a pharmaceutical (tamoxifen) to otherwise healthy women, who are designated to be at high risk of breast cancer (Fosket 2010). Proto-diseases are an expansion of medicalization (Conrad 2007), and the rise of "surveillance medicine" conjures the need for "anticipatory care . . . transform(ing) the future by changing the health attitudes and health behaviours of the present" (Armstrong 1995, 402). Indeed, the medical trend toward anticipatory medicalization impacts other health and social arenas outside clinical medicine; in the case of preconception care, public health campaigns have been designed in order to effect change in health attitudes and health behaviors among women in their everyday lives, with an eye toward preventing adverse reproductive outcomes in the future (Waggoner 2015). Anticipatory medicalization refers to measurable forms of risk (e.g., biomarkers for certain conditions, such as pre-hypertension or pre-diabetes) as well as any demonstrable effects on subjective feelings of risk.

Building on the literature on prediction, predisposition, and proto-disease, the concept of anticipatory medicalization highlights the broad expansion of a preemptive form of medicalization. Anticipatory medicalization is not just limited to preconception care or the arena of reproduction. Any moment in which the processes of medicalization—framing or defining a condition in medical terms—are mobilized in advance of the presence of a condition; any moment in which the prediction or prognosis of a putative condition is defined based on an examination of future, as yet unseen risks, then anticipatory medicalization may be observed.

Anticipatory medicalization is about medicalizing a putative risk that may or may not appear in the future. As a form of prevention, we recognize anticipatory medicalization as having the potential to be efficacious. Just as many forms of medical prevention have potential health benefits, anticipatory medicalization may spur positive health consequences, such as increased patient awareness of healthy behaviors, patient empowerment, and early clinical detection of potential problems. However, regardless of the contributions to prevention, which may be substantial, anticipatory medicalization increases medicalized categories, and this process may have numerous other impacts. Potential consequences of anticipatory medicalization include the increased medicalization of life, wider medical surveillance, the merging of risk and disease (Aronowitz 2009), potential overdiagnosis, potential overtreatment, unnecessary treatment, uncertain prognoses, the expansion of medical jurisdiction, heightened patient anxiety, and conceivable costs related to medical treatment and screening. These variables need to be taken into account when evaluating the costs and benefits of varied forms of predictive medicine. Notwithstanding all potential benefits or ramifications, it is clear that prevention has been medicalized today in a manner that hinges on anticipation.

Note

1 This section is adapted from Conrad 2013.

References

Adams, V., Murphy, M., and A. E. Clarke. 2009. "Anticipation: Technoscience, Life, Affect, Temporality." *Subjectivity* 28:246–65.

Armstrong, D. 1995. "The Rise of Surveillance Medicine." *Sociology of Health & Illness* 17:393–404.

Armstrong, E.M. 2000. "Lessons in Control: Prenatal Education in the Hospital." *Social Problems* 47:583–605.

Armstrong, E.M. 2003. *Conceiving Risk, Bearing Responsibility: Fetal Alcohol Syndrome and the Diagnosis of Moral Disorder*. Baltimore: The Johns Hopkins University Press.

Aronowitz, R.A. 2009. "The Converged Experience of Risk and Disease." *The Milbank Quarterly* 87:417–42.

Atrash, H., B.W. Jack, Johnson, K., et al. 2008. "Where Is the 'W'oman in MCH?" *American Journal of Obstetrics & Gynaecology* (Suppl.):S259–65.

Bandini, J. 2015. "The Medicalization of Bereavement: (Ab)normal Grief in the DSM-5." *Death Studies* 39(6):347–52.

Barker, K.K. 1998. "A Ship Upon a Stormy Sea: The Medicalization of Pregnancy." *Social Science & Medicine* 47:1067–76.

Barker, K.K. 2003. "Birthing and Bureaucratic Women: Needs Talk and the Definitional Legacy of the Sheppard-Towner Act." *Feminist Studies* 29:333–55.

Bell, S.E., and A.E. Figert. 2015. *Reimagining (Bio)medicalization, Pharmaceuticals, and Genetics: Old Critiques and New Engagements*. London: Routledge.

Christakis, N.A. 1999. *Death Foretold: Prophecy and Prognosis in Medical Care*. Chicago: The University of Chicago Press.

Clarke, A.E., Shim, J.K., Mamo, L., Fosket, J.R., and J.R. Fishman. 2003. "Biomedicalization: Technoscientific Transformations of Health, Illness, and US Biomedicine." *American Sociological Review* 68(2):161–94.

Conrad, P. 1975. "The Discovery of Hyperkinesis: Notes on the Medicalization of Deviant Behavior." *Social Problems* 23(1):12–21.

Conrad, P. 2005. "The Shifting Engines of Medicalization." *Journal of Health and Social Behavior* 46(March):3–14.

Conrad, P. 2007. *The Medicalization of Society: On the Transformation of Human Conditions into Treatable Disorders*. Baltimore: The Johns Hopkins University Press.

Conrad, P. 2013. "Medicalization: Changing Contours, Characteristics and Contexts." In *Health Sociology on the Move: New Directions in Theory*, edited by W. Cockerham, 195–214. Oxford: Blackwell.

Conrad, P., and M. Bergey. 2014. "The Impending Globalization of ADHD: Notes on the Expansion and Growth of a Medicalized Disorder." *Social Science and Medicine* 122:31–43.

Conrad, P., and D. Potter. 2000. "From Hyperactive Children to ADHD Adults: Observations on the Expansion of Medicalized Categories." *Social Problems* 47:59–82.

Conrad, P., and D. Potter. 2004. "Human Growth Hormone and the Temptations of Biomedical Enhancement." *Sociology of Health and Illness* 26:184–215.

Fosket, J.R. 2010. "Breast Cancer Risk as Disease: Biomedicalizing Risk." In *Biomedicalization: Technoscience, Health, and Illness in the U.S.*, edited by A.E. Clarke, L. Mamo, J.R. Fosket, J.R. Fishman, and J.K. Shim, 331–52. Durham: Duke University Press.

Gillespie, C. 2012. "The Experience of Risk as 'Measured Vulnerability': Health Screening and Lay Uses of Numerical Risk." *Sociology of Health & Illness* 34(2):194–207.

Hinshaw, S. 1994. *Attention Deficit Disorder and Hyperactivity in Children*. Thousand Oaks: Sage.

Horwitz, A., and J. Wakefield. 2007. *The Loss of Sadness*. New York: Oxford University Press.

Illich, I. 1975. *Medical Nemesis*. New York: Pantheon.

Johnson, K., et al. 2006. "Recommendations to Improve Preconception Health and Health Care—United States: A Report of the CDC/ATSDR Preconception Care Work Group and the Select Panel on Preconception Care." *MMWR* 55(RR06):1–23.

Konrad, M. 2003. "Predictive Genetic Testing and the Making of the Pre-Symptomatic Person: Prognostic Moralities amongst Huntington's-Affected Families." *Anthropology & Medicine* 10:23–49.

Lane, C. 2007. *Shyness: How Normal Behavior Became a Sickness*. New Haven: Yale University Press.

Loe, M. 2004. *The Rise of Viagra*. New York: NYU Press.

Rosenberg, C. E. 2007. "Banishing Risk: Or, the More Things Change, the More they Remain the Same." In *Our Present Complaint: American Medicine, Then and Now*, 60–76. Baltimore: The Johns Hopkins University Press.

Rosenberg, C. E. 2009. "Managed Fear." *The Lancet* 373:802–3.

Rothman, B. K. 1993. *The Tentative Pregnancy: How Amniocentesis Changes the Experience of Motherhood*. New York: W.W. Norton & Company.

Schwarz, A., and S. Cohen. 2013. "ADHD Seen in 11% of U.S. Children as Diagnoses Rise." *New York Times*, March 31.

Waggoner, M. R. 2013. "Motherhood Preconceived: The Emergence of the Preconception Health and Health Care Initiative." *Journal of Health Politics, Policy and Law* 38:345–71.

Waggoner, M. R. 2015. "Cultivating the Maternal Future: Public Health and Pre-Pregnant Self." *Signs: Journal of Women in Culture and Society* 40(4):939–62.

Wertz, R. W., and D. C. Wertz. 1977. *Lying-in: A History of Childbirth in America*. New Haven: Yale University Press.

Zola, I. K. 1972. "Medicine as an Institution of Social Control." *Sociological Review* 20(4):487–504.

8 Predicting the Cost of Diseases in Resource-Poor Countries

Steffen Flessa

Interventions against diseases almost always have a long-term impact that calls for the prediction of epidemiological and economic parameters. The consequences of interventions in the future determine the decisions of health policy makers of today. Thus, predicting the cost of diseases is crucial for public health planning.

This paper focuses on AIDS, malaria, diabetes, and cervical cancer as examples of diseases which cause tremendous human suffering, kill millions every year, and challenge health care systems in developing as well as in developed countries. As health care resources are notoriously scarce in these regions, the wrong decisions or misallocation of funds will directly result in human increased morbidity and mortality. However, "right" or "wrong" depends strongly on the value of future health gains and costs of interventions. Following the introduction, this chapter will give an overview of different possibilities of predicting the future in health economic models. Afterwards, it will focus on four examples, i.e., AIDS, malaria, diabetes, and cervical cancer.

Introduction

In order to appreciate the relevance of predicting the cost of diseases in developing countries, it is necessary to understand the context of the predictive models in health economics. Generally, economics is the science of explaining and solving the problem of scarcity by efficiency (Flessa and Greiner 2013). We face a lot of scarcity, such as scarcity of materials, equipment, buildings, staff, motivation, ideas, and health. In resource-poor countries, scarcity is even worse so that economics is even more important in these regions (Jack 1999). Efficiency is the deep core of economics and simply means that the quotient of results and inputs is maximized. This can be done by maximizing the output at a given input, or by minimizing the input at a given output.

$$E = \frac{Results}{Inputs} = \frac{\sum_{i=1}^{n} w_i x_i}{\sum_{j=1}^{m} v_j y_j} \rightarrow Max!$$

with

E Efficiency
w_i Weight of output i
x_i output i
v_j Weight of input j
y_j input j
m number of inputs
n number of outputs

In reality, however, we face a number of problems in measuring and evaluating results and inputs. One major input problem is that they occur in different periods, for instance today, next year, or in twenty years. Economists solve this problem by discounting the value of future inputs to a present value. As inputs are frequently expressed in currency units, the interest rate is an appropriate discounting rate (Schöffski and Schulenburg 1998).

$$NC = \sum_{t=0}^{T} C_t \left(1 + \frac{r}{100}\right)^{-t}$$ with

NC Net value of cost
T Time horizon
C_t Cost in period t
r interest rate [%]

The same is true for the result dimension of the efficiency formula. Results occur in different periods and have to be discounted. Here discounting expresses the time preference as the systematic devaluation of future utilities in comparison to present utilities (Muennig 2002).

However, the time preference rate or the rate of discounting future results is under dispute (Viscusi and Moore 1989; Parsonage and Neuburger 1992; Krahn and Gafni 1993; Lipscomb et al. 1996; Walker and Kumaranayake 2002). The higher the discounting rate, the lower the value of future results for the decision. Some even argue that only today is relevant, or as Morley states in the title of his famous book: "My name is today" (Morley and Lovel 1986). It is not tomorrow that counts for the survival of a child, but today!

The discussion of investments in the health of the present versus the future generation runs like a red thread through the entire health policy arena in resource-poor countries. This discussion is existential as the resources are so limited: Any allocation of funds for prevention will help future generations to survive. But it will also mean that the current generation will have fewer funds for treatment. "Who survives?" is the cruel but realistic question underpinning inter-temporal resource allocation (Fuchs 2011). It calls for consequential ethics considering the long-term consequences of our resource allocation. A pure focus on deontological ethics is likely to result in more suffering and death in the long-run.

Health economists hope to improve the health of the population by making the best possible use of limited resources. Among a number of other issues they have to address is the question whether we should invest in the health of people today or tomorrow. This has an ethical dimension, as all resources which we invest in this generation cannot be used for the future generation and vice versa. All resources that are misallocated are lost to improving the lives of people.

Consequently, we will present four examples of inter-temporal resource allocation and the epidemiological and economic impact. Firstly, however, we would like briefly to define the term "resource-poor country." Here we refer primarily to those currently forty-nine countries which the UN calls "least developed countries." The majority of them are located in Sub-Saharan Africa.

Sub-Saharan Africa is indeed the region with the highest burden of disease per capita (WHO 2008). Life expectancy is much lower than in all other continents while mortality rates and loss of quality of life are much higher. In these countries, health policy making of necessity involves cruel choices. It is a question of allocating life and death between different regions, social groups, and the current and future generations. Consequently, there is a need for heuristics to act in the best possible way in a given situation. These heuristics must not be arbitrary, but based on professional reflection as well as on sound data.

Predictive Models

The prediction of future results and costs is standard in economic evaluations (Schöffski and Schulenburg 1998). For health economics, it regularly means that we have to predict demographic and epidemiological results as well as the costs of an intervention. This is done using a wide variety of models, such as Markov Chains, System Dynamics Models, Bio-metrical Models, or Discrete Event Simulations (DES). Bio-metrical models are rather restricted in their application due to their mathematical structure, and are consequently hardly used in health economic predictions. DES are highly complex. They are used mainly for early epidemics or agent-based systems (e.g., mother-to-child transmission of HIV) (Rauner et al. 2005).

Markov Models define certain conditions, and the probability of transit from one condition to another in a certain period of time (Beck and Pauker 1983). For instance, typical stages could be "non-infected," "infected," and "dead." In reality, Markov Models can have thousands of conditions (see Figure 8.1).

The matrix of transition probabilities can be used to calculate the number of individuals in a certain condition after a number of periods. This is most relevant for chronic-degenerative diseases, where transition probabilities (a_{ij}) remain stable over time. Here the quality of prediction is quite high.

$$\underline{w}'_{t+1} = \underline{w}'_t \bullet A$$

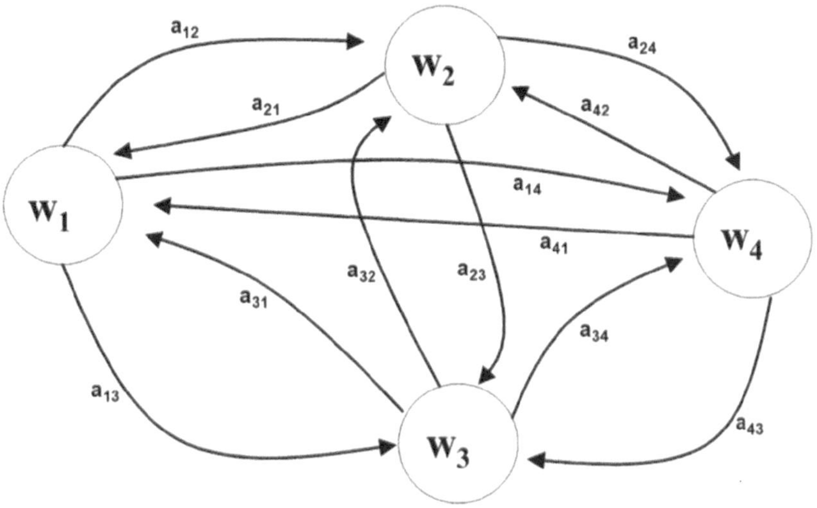

Figure 8.1 Markov Graph.
Source: Flessa and Greiner (2013).

$$\underline{w} = \begin{pmatrix} w_1 \\ \ldots \\ w_n \end{pmatrix}; \; A = \begin{pmatrix} a_{11} & a_{12} & \cdots & a_{1n} \\ a_{21} & a_{22} & \cdots & a_{2n} \\ \vdots & \vdots & \ddots & \vdots \\ a_{n1} & a_{n2} & \cdots & a_{nn} \end{pmatrix}$$

$$\underline{w}'_t = \underline{w}'_0 \bullet A^t \text{ with}$$

A Matrix of transition probabilities
a_{ij} Transition probability from condition i to condition j within one period
w_t Vector of conditions in period t

For infectious diseases, System Dynamics Models are more appropriate as they include feedback loops (Sterman 2000). For instance, the number of eggs which an anopheles mosquito can lay on a certain day depends on the number of anopheles on that day, which itself is a consequence of the number of eggs laid some time before. These feedback loops require complex models, which can have thousands of simultaneous equations (see Figure 8.2) and will require a huge number of scenarios to cope with structural and parameter uncertainty.

The health economist will chose the most appropriate model based on a number of decision variables. Firstly, he has to consider the purpose of the economic analysis. If we want to reflect the epidemiological situation as closely as possible, we will most likely chose a Discrete Event Simulation or

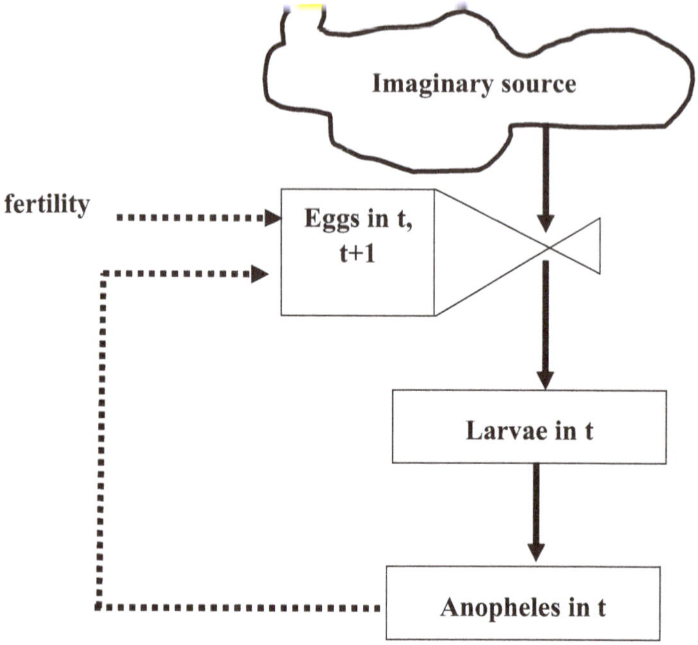

Figure 8.2 System Dynamics of Anopheles.
Source: Flessa (1999).

a System Dynamics Model. For a budget impact analysis, a simpler model could suffice. Secondly, the model will depend on the modelling resources. Discrete Event Simulations and System Dynamics Models are complex and require computer programming. If a Markov Model is sufficient, one should not invest more resources in a System Dynamics Model. Thirdly, infectious diseases are usually more complex than chronic-degenerative diseases, as they frequently require modelling the entire ecology of a host (such as an anopheles), including geographical conditions. Finally, the model will reflect the relevance of the time preference. Simple models will not allow us to distinguish different periods of time, whereas realistic models include discounting of future costs and effects. System Dynamics Models as well as Markov Models are always multi-period models.

This sounds very abstract. Therefore, we will present four examples of high public health relevance in resource-poor countries.

AIDS

AIDS is a human tragedy, and it is also an economic catastrophe for the countries concerned. A major focus of all AIDS-related activities is anti-retroviral therapy. In the developed world, Highly Active Anti-Retroviral Therapy (HAART) is standard, but the majority of HIV cases in resource-poor countries still lack access to multi-drug therapy. It serves as a good example of the relevance of different time horizons in decision-making. In the short-term, ART has a strongly positive impact on the life of patients. But in the long-term it entails some risks. The virus might become resistant, in particular, in countries where poor logistics might lead to interruption of drug supply, and where compliance of patients is poor. The knowledge that there is a "cure"—although it definitely is no cure—might make people less cautious, thus spreading the disease further. And the funds invested in ART will be missing elsewhere, a fact which economists call opportunity costs.

Figure 8.3 shows the wonderful short-term consequences of ART. ART means less disease progression, less death, and less suffering. In the long-term, the picture is more complex and ART might have a number of severe side-effects. Does this mean that we have to withhold this treatment from millions of AIDS patients? Definitely not! But it calls for a thorough analysis of long-term effects, and a prediction of the future epidemiological and economic consequences of public health interventions. This is not possible without modelling.

Figure 8.4 shows the estimated number of HIV and AIDS cases in Tanzania up to 2020. It is based on a multi-compartment System Dynamics Model specifically made for this country in the year 2000. It is obvious that the worst effects are still to come.

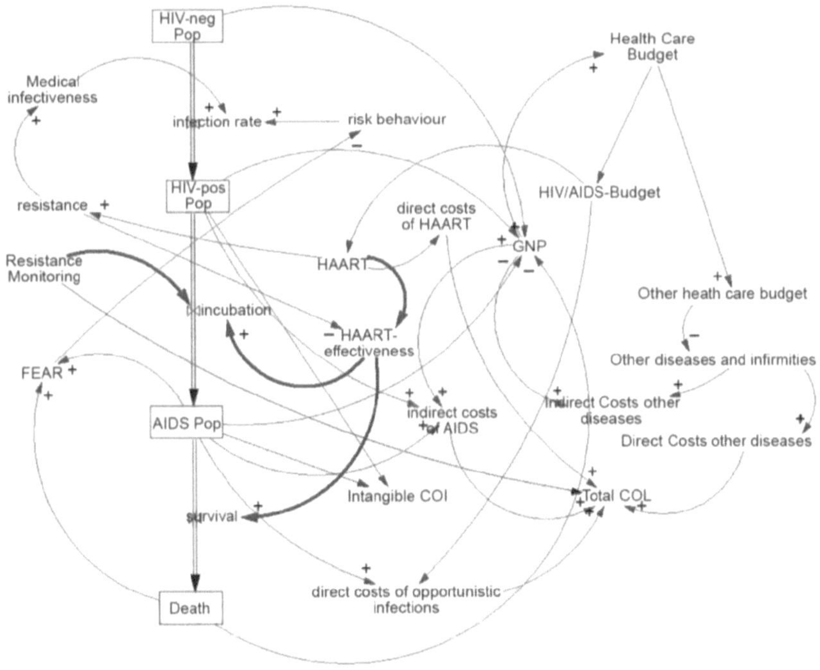

Figure 8.3 System Model of ART.

Source: Flessa (2012).[1]

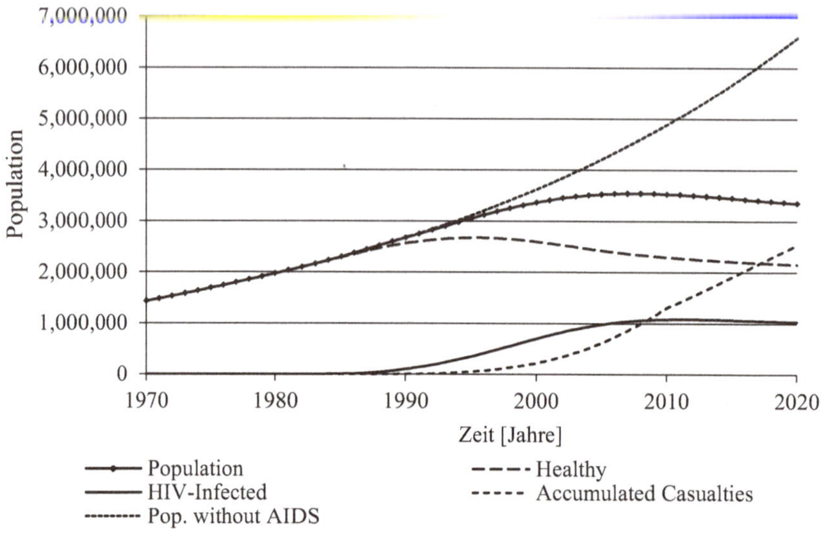

Figure 8.4 Aids in Tanzania.

Source: Flessa (2002).

The corresponding costs of treating opportunistic diseases are enormous. From the year 2000 to the year 2020, the annual treatment cost will go up from US$45 to 120 million, i.e., some US$5 per healthy Tanzanian. This might not sound much to European ears. But in some health districts of Tanzania, the government cannot spend more than US$15 per annum per capita for health care. AIDS is an economic tragedy, and the worst effects are still to come. We need a precise prediction of future developments for budgeting.

Figure 8.5 shows the developments if a vaccination had been introduced in the year 2000. Most important is the scenario "Short" if the vaccination protects for only five years. We realize that short-term vaccination protection will only have short-term effects, and the number of cases will finally increase dramatically. This model suggests strongly that interventions against infectious diseases must be long-term.

Although this prediction model is already some years old, it clearly shows that a disease with an incubation period of ten years calls for the explicit consideration of future effects and costs. This can only be done with dynamic models. The simulation also indicates that any intervention must be long-term, but in reality funding is usually short-term. Most health-related development programs have a time-horizon of three to five years, so that some interventions might be evaluated as cost-effective which in the long-term might have only very limited cost-effectiveness. In other words, cost-effectiveness depends on the time-horizon.

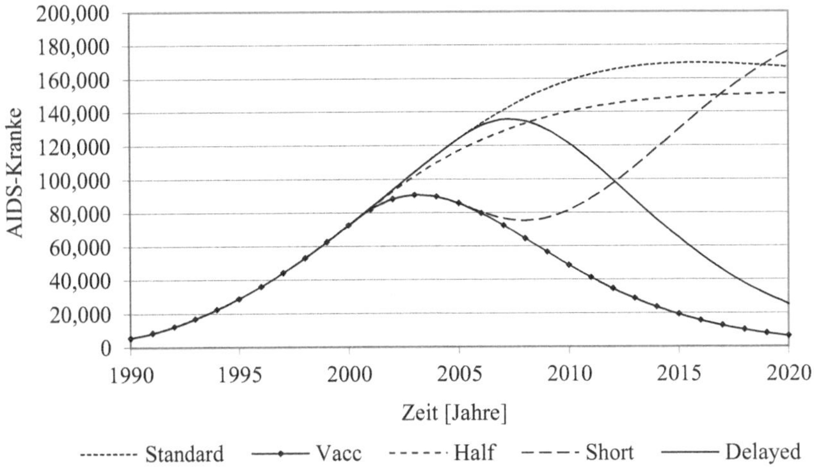

Figure 8.5 Vaccination Scenarios.

Source: Flessa (2002).

Malaria

The second example is Malaria Tropica—a disease that results in some 216 million cases and 655,000 deaths p.a., 91% of which are in Africa. The parasite is transmitted by the anopheles mosquito.

Malaria Tropica is a disease of high economic relevance. Costing studies in Kenya show that the treatment of one case of simple malaria costs around US$5; for complicated malaria the average costs are US$30 (Flessa et al. 2011). However, these are only the treatment costs. Patients will lose on average ten working days due to a malaria attack. Economically also relevant is the fact that malaria is in many places a seasonal disease requiring relevant excess capacity in the health care facilities. Other major cost components are the preventions programs. WHO has invested huge amounts to eradicate the disease, but finally had to give up. Today we try to roll back malaria by limited in-door spraying, control of breeding places, and the use of insecticide treated bed-nets, in particular for infants.

Again we developed a System Dynamics Model to predict the impact of interventions. Figure 8.6 shows the consequences of in-door spraying to kill the mosquito after the blood-meal. The Figure shows that this intervention might have a high impact on the number of cases in the short run. However, the anopheles population will recover unless the budget is increased at least proportionally to the growth of the human population. Any short-term analysis will be misleading. Unfortunately, most evaluation reports in International Public Health have a two to three year perspective.

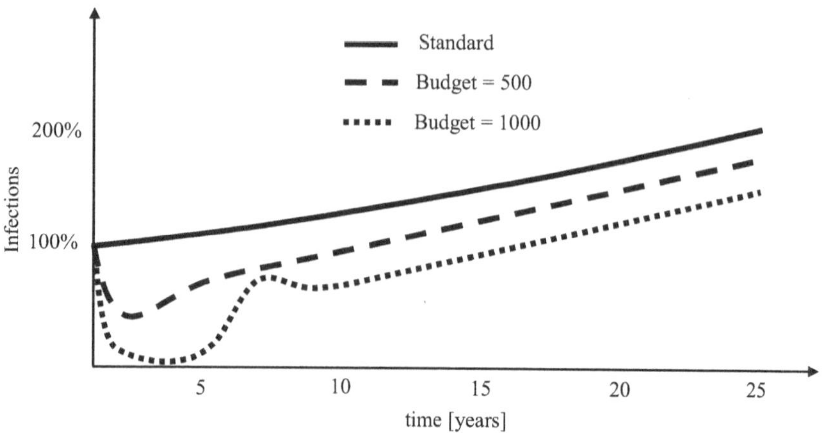

Figure 8.6 In-door Spraying.
Source: Flessa (2002).

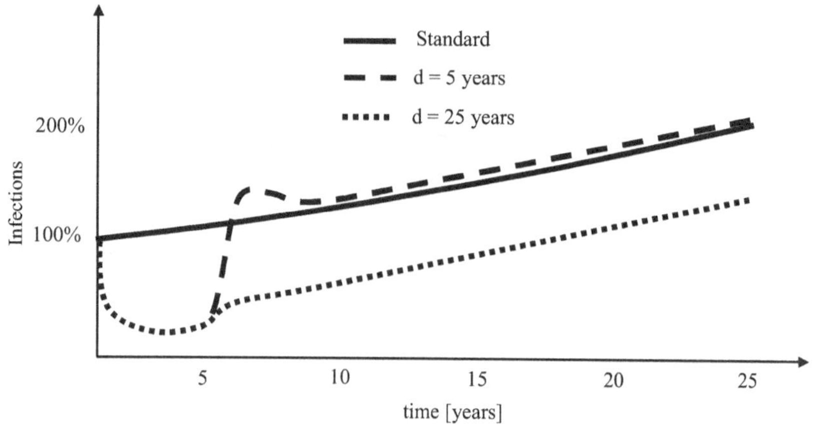

Figure 8.7 Bed-net Programs.
Source: Flessa (2002).

Figure 8.7 shows the impact of a bed-net program. We see the strong benefit of bed-nets if all children are protected. However, nets will have to be replaced at least every five years. If this does not happen, we will see a tremendous come-back of malaria. Those who had been protected before have not developed semi-immunity and will become "easy" victims of malaria once the bed-net is gone. We either finance bed-net programs for a longer time or we have to face catastrophic consequences.

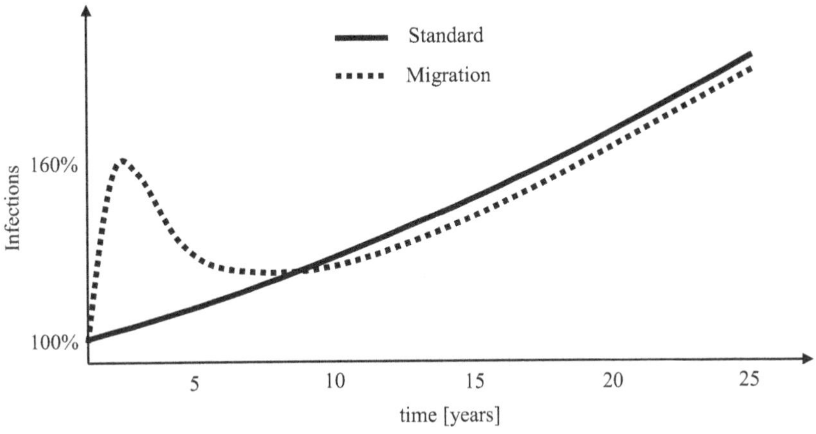

Figure 8.8 Migration.
Source: Flessa (2002).

Finally, some countries have transferred parts of their population from highly populated areas to other places (Figure 8.8). Frequently, they were transferred from malaria-free mountain areas to swampy places full of anopheles. The consequence, for example of the transmigrasi project in Indonesia, was a catastrophic malaria epidemic for several years. Every development intervention must consider health impacts: short- and long-term.

Based on these predictions we can learn that Roll-Back Malaria projects need sustainable interventions and long-term financing. Furthermore, whoever wants to help these countries in their economic development must consider long-term effects of agricultural, social, and economic programs. This will involve the regular prediction of epidemiological and economic effects of interventions.

Diabetes

The third example is Diabetes mellitus type II. The International Diabetes Federation (IDF) estimates that some 350 million people are affected worldwide, and 90% of diabetes patients live in low- or middle-income countries (IDF 2014). This sounds surprising, as we still consider people in developing countries to be starving. But the reality is different, as in particular South East Asia is facing a diabetes epidemic causing extreme human suffering and costs (WHO 2013; Flessa and Zembok 2014).

In the year 2013, we were asked by the Ministry of Health of Cambodia to support the implementation of the new policy to combat non-communicable diseases. For that purpose, we developed a Markov Model that allows us to predict the number of cases and the costs of treatment as well as the economic assessment of interventions (Flessa and Zembok 2014). Currently, it is estimated that about 2.9% of the adult population has diabetes. Figure 8.9 exhibits the predicted development of diagnosed and undiagnosed cases as a foundation of the analysis of future relevance of this disease.

Most relevant for the ministry is the fact that a majority of diagnosed cases will have complications and require professional care. Until now diabetes is not part of the basic package of health care interventions in Cambodia. Thus, no public funds are allocated to prevent and treat diabetes. Only the rich minority can afford drugs and insulin.

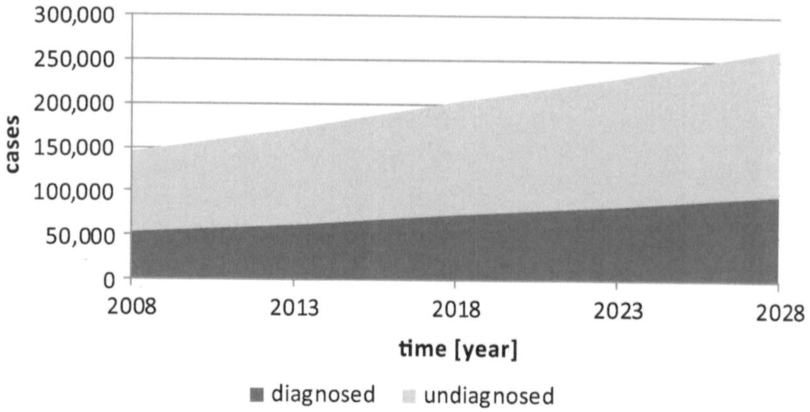

Figure 8.9 Diabetes: Diagnosed and undiagnosed cases.
Source: Flessa and Zembok (2014).

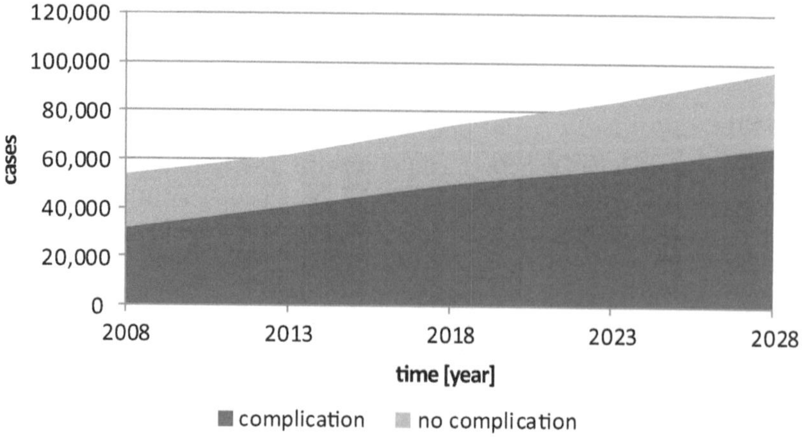

Figure 8.10 Diabetes: Diagnosed cases with and without complications.
Source: Flessa and Zembok (2014).

Figure 8.11 shows that the majority of patients will require anti-diabetic drugs, but these drugs are not available in public health centers. Consequently, effort has to be invested to improve logistics and financing for this medication.

For the ministry it is crucial to have an estimate of how much professional diabetes care would cost. Therefore, we calculated the budget impact. Again, an amount of US$4 million in the year 2028 might sound negligible to most Europeans, but for a least developed country like Cambodia, this is a severe burden (Figure 8.12).

Finally, we calculated the impact of improved access to drugs and insulin. Figure 8.13 shows that improved access to drugs will result in a lower

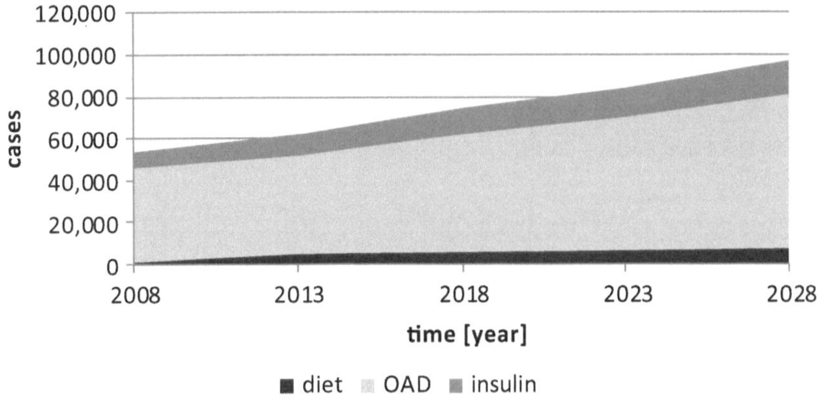

Figure 8.11 Diabetes: Therapy of diagnosed cases.

Source: Flessa and Zembok (2014).

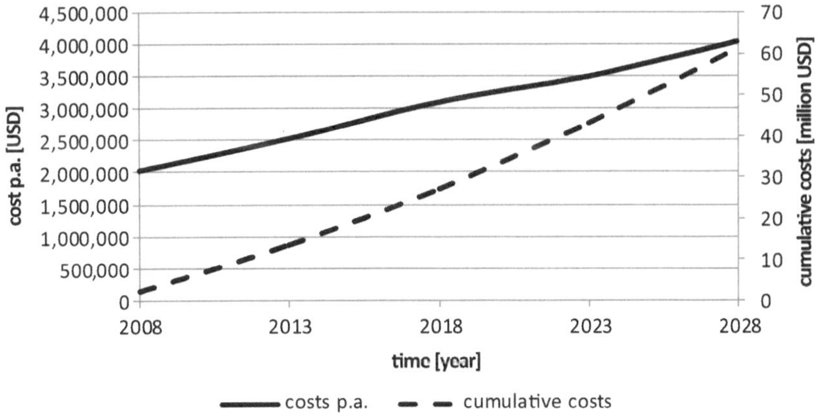

Figure 8.12 Diabetes: Budget impact.

Source: Flessa and Zembok (2014).

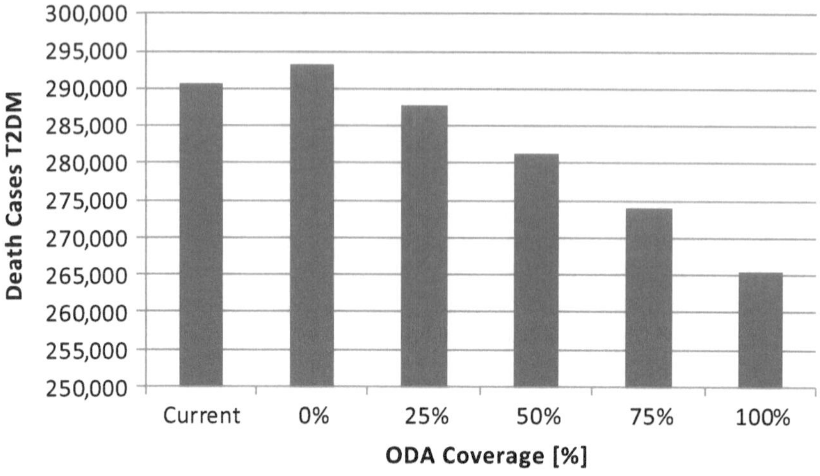

Figure 8.13 Diabetes: Impact of OAD coverage (current = 12.5%).
Source: Flessa and Zembok (2014).

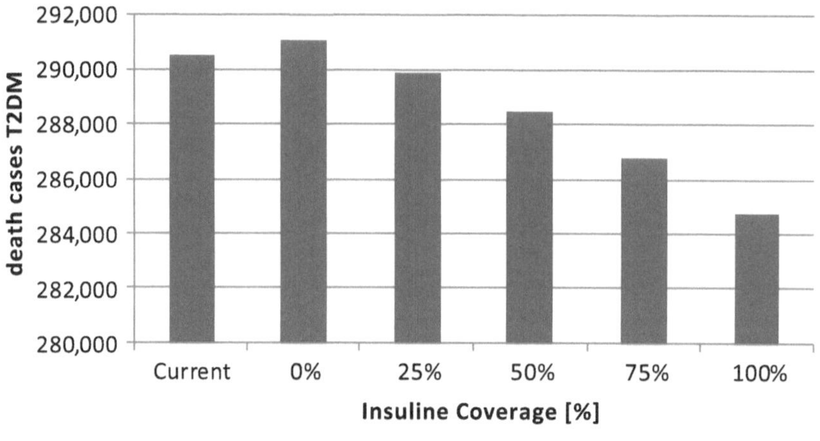

Figure 8.14 Diabetes: Impact of insulin coverage (basic = 12.5%).
Source: Flessa and Zembok (2014).

number of diabetes-related deaths. The intervention is highly cost-effective as the Incremental Cost per Life Saved is US$800. This amount includes discounting costs and effects with a rate of 5%.

Fighting mortality by provision of insulin is not as cost-effective (Figure 8.14), but with Incremental Costs per Life Saved of US$3,392, it is still recommendable. The only problem in Cambodia is not everything that is cost-effective can also be financed through local funds. In other words, the ability to pay is less than the willingness to pay.

Diabetes is "the sweet epidemic" and it will increase strongly in many low- and middle-income countries with the rapidly aging societies. This means that we have to plan now in order to budget for this future epidemic. Interventions to combat the disease should be cost-effective. Thus, diabetes is a paradigmatic example for other non-communicable diseases. We need sustainable interventions and long-term financing. And we need health economic models to predict the epidemiological and economic effects of interventions.

Cervical Cancer

The last example is cervical cancer—a cancer that is caused by the human papillomavirus (HPV). The WHO estimates that some 500,000 new cases occur annually and some 270,000 women die from it (WHO 2014b). Eighty percent of deaths occur in low- and middle-income countries (WHO 2014a). As the Ministry of Health of Cambodia included this disease in its NCD policy, we were asked to assess the cost-effectiveness of different interventions (Flessa and Dietz 2014).

Figure 8.15 shows the impact of interventions on the number of death cases in Cambodia. It is obvious that treatment, screening, and vaccination have positive impacts.

However, the costs of these interventions are also quite high. The most expensive single intervention is a vaccination program (Figure 8.16).

Furthermore, the fruits of such a vaccination program will be harvested only after decades. HPV-vaccination is fully effective if girls are vaccinated

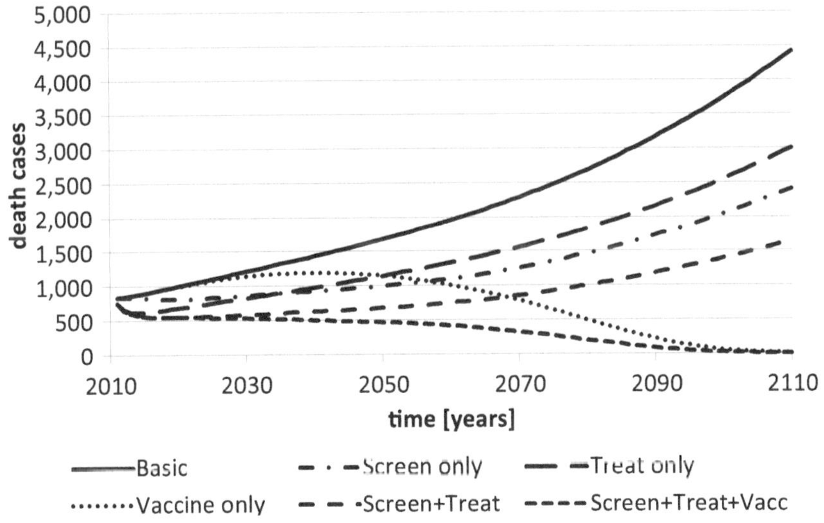

Figure 8.15 Cervical Cancer: Combined intervention.

Source: Flessa and Dietz (2016).

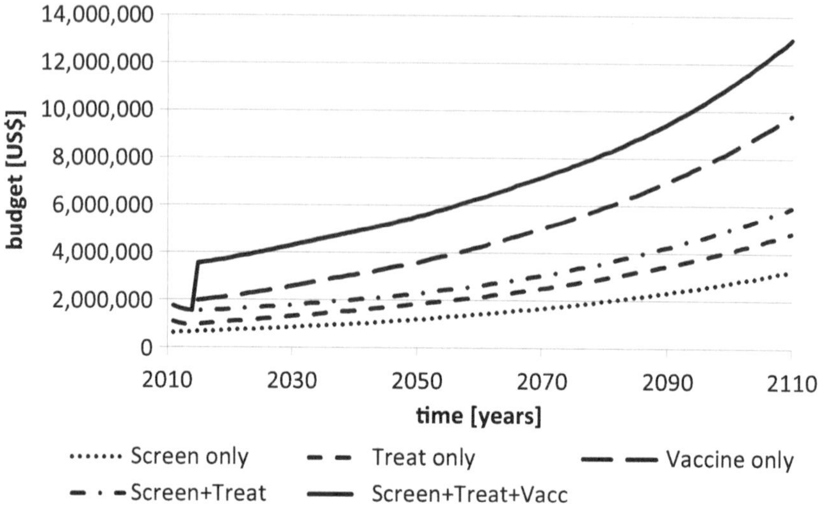

Figure 8.16 Cervical Cancer: Budget.

Source: Flessa and Dietz (2016).

before their first sexual intercourse. WHO recommends vaccination before the age of 15 years. The infection happens on average at 25 years. The first lesions appear on average at 41 years, and invasive cancer is diagnosed at 49 years. On average, a woman dies at the age of 51 from cervical cancer. This is thirty-six years after a possible vaccination. Thus, the time horizon and the time preference are of highest relevance for the decision-making on the most appropriate intervention strategy in Cambodia.

Figure 8.17 shows the conditions under which different interventions are cost-effective. All combinations of time horizon and discounting rate to the lower right of the respective curve are cost-effective. It is obvious that a higher time horizon makes it more likely that an intervention is cost-effective. In particular, a stand-alone vaccination program requires a time horizon of at least 55 years. Otherwise this intervention will not be considered in public policy irrespective of the discounting rate. In Cambodia, the majority of people struggle to survive until the next harvest. An intervention that will pay back only after 55 years is irrelevant. Treatment, instead, will be cost-effective if the policy maker considers at least the next ten years as relevant. This statement is true for all realistic discounting rates. Thus, predicting the future costs and cost-effectiveness of interventions requires an assessment of the time horizon and the time preference of a particular culture.

In summary, we can conclude that health care in resource-poor countries always entails an existential dimension. Inter-regional, inter-sectorial, and inter-temporal resource allocation is always a life or death decision (Fuchs 2011), not only on better quality of life. "Should we spend the same funds

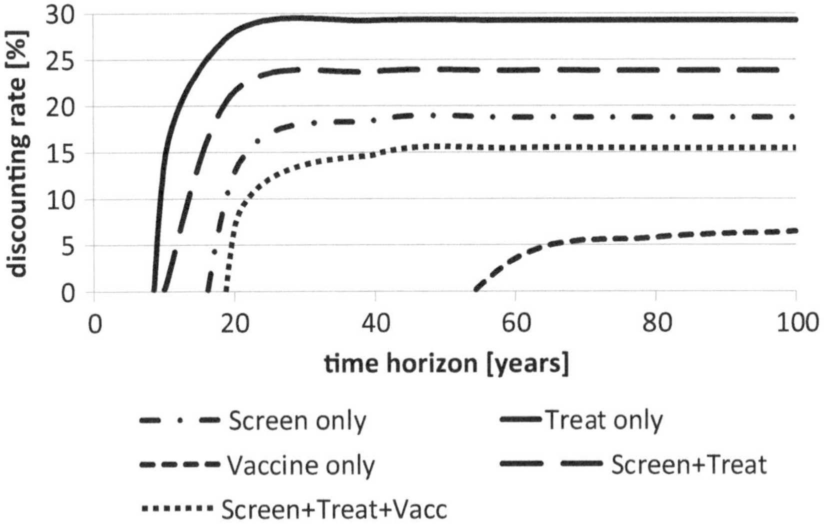

Figure 8.17 Cervical Cancer: Role of time horizon and discounting.

Source: Flessa and Dietz (2014).

to treat one patient with chronic kidney disease, or 1,000 children with diarrhea?" "Should we spend the same funds to care for the urban minorities, or for the rural majorities without political influence?" "Should we invest in vaccination programs that will save the lives of millions in twenty years, or treat the patients of today?" These questions lie at the core of health economics—and they are definitely questions of an ethical dimension. These questions call for an answer based on precise calculations of health economic models. This is the utmost purpose of health economics: to support decision-making and make the best possible use of scarce resources. Or as we would like to put it: Health Economics is there to serve life—*Oeconomia ancilla vitae*!

Note

1 COI: Cost of Illness; GNP: Gross national product; signs indicate relation between root and arrow (e.g., if direct cost of other diseases increase, then the total cost of illness will increase: +; if resistance increases, HAART effectiveness will decrease).

References

Beck, J.R., and S.G. Pauker. 1983. "The Markov Process in Medical Prognosis." *Medical Decision Making* 3:419–58.

Flessa, S. 1999. "Decision Support for Malaria-Control Programmes—a System Dynamics Model." *Health Care Management Science* 2:181–91.

Flessa, S. 2002. *Malaria und Aids: Gesundheitsökonomische Analysen auf Grundlage von Disease Dynamics Modellen*. Lage: Jacobs Verlag.

Flessa, S. 2012. Internationales Gesundheitsmanagement. Oldenbourg, München, page 139.

Flessa, S., Dietz, D., and Weiderpass, E. 2016. Health policy support under extreme uncertainty: the case of cervical cancer in Cambodia. *EURO Journal on Decision Processes*, Vol. S. 183–218.

Flessa, S., and W. Greiner. 2013. *Grundlagen der Gesundheitsökonomie*. Berlin/Heidelberg: Springer Gabler.

Flessa, S., Moeller, M., Ensor, T., and K. Hornetz. 2011. "Basing Care Reforms on Evidence: The Kenya Health Sector Costing Model." *BMC Health Services Research* 11:128.

Flessa, S., and A. Zembok. 2014. "Costing of Diabetes Mellitus Type II in Cambodia." *Health Economics Review* 4:1–15.

Fuchs, V. R. 2011. *Who Shall Live? Health, Economics and Social Change*. Hackensack: World Scientific.

IDF. 2014. "IDF Diabetes Atlas." Available at http://www.idf.org/diabetesatlas/FAQs [accessed February 2, 2014].

Jack, W. 1999. *Principles of Health Economics for Developing Countries*. Washington: The World Bank.

Krahn, M., and A. Gafni. 1993. "Discounting in the Economic Evaluation of Health Care Interventions." *Medical Care* 31:403–18.

Lipscomb, J., Weinstein, M.C., and G.W. Torrance. 1996. "Time Preference." In *Cost-Effectiveness in Health and Medicine*, edited by M.R. Gold, J. E. Siegel, L. B. Russell, and M. C. Weinstein, 214–46. New York: Oxford University Press.

Morley, D., and H. Lovel. 1986. *My Name Is Today*. London: Blackwell.

Muennig, P. 2002. *Designing and Conducting Cost-Effectiveness Analysis in Medicine and Health Care*. San Francisco: Jossey-Bass Inc.

Parsonage, M., and H. Neuburger. 1992. "Discounting and Health Benefits." *Health Economics* 1:71–6.

Rauner, M.S., Brailsford, S.C., and S. Flessa. 2005. "The Use of Discrete-Event Simulation to Evaluate Strategies for the Prevention of Mother-to-Child Transmission of HIV in Developing Countries." *Journal of the Operational Research Society* 56:222–33.

Schöffski, O., and J.-M.G.V.D. Schulenburg. 1998. *Gesundheitsökonomische Evaluationen*. Berlin: Springer Verlag.

Sterman, J. 2000. *Business Dynamics*. Boston: McGraw-Hill.

Viscusi, W.K., and M.J. Moore. 1989. "Rates of Time Preference and Valuations of the Duration of Life." *Journal of Public Economics* 38(3):297–317.

Walker, D., and L. Kumaranayake. 2002. "Allowing for Differential Timing in Cost Analyses: Discounting and Annualization." *Health Policy Plan* 17:112–8.

World Health Organization (WHO). 2008. *Global Burden of Disease Update 2004*. Genf: Weltgesundheitsorganisation.

World Health Organization (WHO). 2013. *Diabetes*. Available at http://www.who.int/mediacentre/factsheets/fs312/en/ [accessed February 24, 2014].

World Health Organization (WHO). 2014a. *GLOBOCAN*. Geneva: World Health Organisation. Available at http://globocan.iarc.fr/Pages/fact_sheets_cancer.aspx [accessed January 23, 2014].

World Health Organization (WHO). 2014b. *Health Statistics and Health Information Systems*. Geneva: WHO. Available at http://www.who.int/healthinfo/global_burden_disease/estimates_regional/en/index.html [accessed January 2, 2014].

9 Genetic Disorders in Chinese Patients and Their Families

A Call for Action on Predictive Medicine

Xian-Ning Zhang and Ji Zuo

Introduction

The recent history of genetics in relation to medicine has been one of breathtaking discovery, and already patients and families are benefiting (Hudson 2011). With the enormous achievement of the human genome project (HGP), molecular biology, and bioinformatics, genomic medicine has been suffusing the mainstream of medical science and has claimed a leading role in precision medicine—to prevent, diagnose, screen, and treat cancers, inherited genetic disorders, infectious diseases, as well as many other diseases and environmental influences reliably and effectively (Collins and Varmus 2015; Lander 2015). Even though there is no reliable data from the well-organized national survey, the total incidence of genetic disorders per year in China is likely to be high (Huang and Qi 2005; Lo 2008; Dai et al. 2011; Li 2013). According to the estimated frequency (~1%) proposed by Carter (1977), there are more than 16,800,000 individuals estimated to be affected by the common monogenetic disorders nationwide. With regard to Chinese patients' and their families' hereditary conditions, it is time to emphasize the possibilities that genomic technologies can offer to families-at-risk in the prediction and prevention of genetic diseases. It is crucial to recognize the important role of "*yousheng*" in China (loosely and somewhat misleadingly translated as "eugenic science" in English), which means "to give birth to a healthy baby" (Li 1980; Wu 1984, 155; Chen et al. 1999; Wang 2004). The concept of "*yousheng*" is widely understood and accepted by the Chinese people. In this article, where necessary, we employ the term "eugenics" as the conventional translation of "*yousheng*," but we use the term advisedly, as is explained in detail below.

When Chinese talk about "quality of population," this should not be taken to imply a certain race or ethnic group. Nor is it the aim of Chinese geneticists to clean the gene pool by precluding "birth defects" (Dickson 1998; Chen et al. 1999; Wang 2004). The public in general hopes to improve the general health of the Chinese population by decreasing the number of deleterious genetic disorders in the population (Dickson 1998; Chen et al. 1999). However, the decision to bear a handicapped child must remain the inviolable right of the parents and not be subject to coercion, intimidation, or force. Few scholars would confuse the liberal technologies and practices of prenatal testing, or preimplantation genetic diagnosis (PGD), with those of forced sterilization, race-based immigration restriction, or in Nazi Germany, the murder of children with disabilities, the killing of mental patients, or the "Lebensborn" program (Paul 2014). Therefore, one should distinguish between eugenics in the latter and former senses: "The old eugenics was genetics and the new genetics is eugenics" (Ekberg 2007; Paul 2014).

A Brief History of *Yousheng* in China

There are many different definitions of the term "eugenics" in the western literature on the subject. In the field of human and medical genetics, eugenics is the science or study of the genetic and prenatal influences that affect expression of certain characteristics in offspring (Garver and Garver 1991; Wynbrandt and Ludman 2008, 150). In contrast to the discipline of genetics, eugenics additionally includes a set of assessments (such as "advantageous/disadvantageous") and practices in order to promote/prevent certain characteristics. The term "eugenics" was originally coined by the English scientist Francis Galton (1822–1911) in 1883 (Bashford and Levine 2010). In Greek it means "well-born" or "noble in heredity" (Garver and Garver 1991; Paul 2014). Galton described eugenics as "the study of agencies under social control that may improve or impair the racial qualities of future generations either physically or mentally" (Bashford and Levine 2010). In fact, advocacy of eugenics—the control of breeding in the service of improving the human race—extends back at least to Plato (429?—347 B.C.) and Aristotle (384–22 B.C.) (Paul 2014). Eugenics, like genetics itself, has had many variants internationally, and has attracted opponents and critics from the moment of its emergence (Paul and Spencer 1995; Bashford and Levine 2010; Allen 2011). While the term eugenics was used very widely, the varieties of policies, practices, sciences, and ideologies conducted in its name defy any simple generalizations (Garver and Garver 1991; Paul 2014). After World War II, when people hear the word "eugenics," the images provoked are of compulsory sterilization of the so-called "feebleminded," immigration restriction, and other racist legislation (Garver and Garver 1991; Dickson 1998; Bashford and Levine 2010; Paul 2014). The workshop on the science and ethics of eugenics held during the 18th International Congress on Genetics in Beijing on August 10–15, 1998 even concluded that "the term 'eugenics' is used in so many different ways as to make it no longer suitable for use in scientific literature" (Dickson 1998). However, in China, "eugenics" in the less controversial sense of "*yousheng*" referred to above is a term applied to preventive genetic medicine; i.e., the aim of helping patients and their families to manage the pain and suffering caused by genetic diseases and reduce the incidence of genetic diseases or the frequency of alleles considered deleterious in the population (Chen et al., 1999; Wang 2004). For example, in the *Chinese Terms in General Practice* (2014) and *Community*

Health and the Chinese Terms in Family Planning (2016) delivered by the government organization, the China National Committee for Terms in Sciences and Technologies, the former defines "*yousheng*" (corresponding to the English term "eugenics"):

> In the narrow sense *yousheng* denotes the reduction of hereditary disorders at birth and congenital anomalies through medical practices; in the broad sense *yousheng* denotes the avoidance of fetal exposure to harmful environment factors during the pre-pregnancy and gestational period in order to bear a healthy offspring.

The latter defines "*yousheng*" (corresponding to the English term "aristogenics" or "improvement in prenatal care") as: "Avoidance of the risk of birth defects and ensuring healthier offspring through health education and medical examination procedures" (China National Committee for Terms in Sciences and Technologies, *www.termonline.cn*). If the history of eugenics cannot provide straightforward lessons for policy, it does teach us that neither moralism nor complacency is justified (Paul and Spencer 1995; Ekberg 2007; Paul 2014; Joly et al. 2016; Paul 2016).

In ancient China, Chinese people had already observed and recognized the importance of "*yousheng*" for human health. For example, *ZuoZhuan* (*Tso Chuan*, *Chronicle of Zuo, Commentary of Zuo*, or *Zuo Tradition*), one of the earliest Chinese books of narrative history covering the period from 722 to 468 B.C., recorded that "if two persons with the same family name get married, their children will be reproductively sterile and infertile" (Li 1980; Wu 1983; Wu 1984, 155). Though people knew nothing about a "gene" or "expression" at that time, a majority of Chinese geneticists see this old description as an indication of a basic genetic principle: that the mating of two individuals with recent common ancestors is often associated with an increased risk of genetic disorders or birth defects in offspring. This is due to the potential for expression of aberrant recessive hereditary characteristics and a slightly increased risk of multifactorial conditions requiring a combination of several genes for expression. Consequently, consanguineous marriage was not encouraged.

After the Opium Wars in 1840, China suffered a series of defeats in international military campaigns and was forced to sign unequal treaties, pay enormous indemnities, lease out treaty ports as well as mining, railway, and water privileges, and even cede Hong Kong, Macau, and Taiwan to Great Britain, Portugal, and Japan. Thus, from the time of the 1890s, *yousheng* became an important element in Chinese political reforms allowing a critique of imperialist encroachment while offering a program for improving and strengthening the nation (Wu 1984, 155; Chung 2010; Sihn 2010). *Yousheng* developed when Pan Guang-Dan (known as Quentin Pan, 1899–1967), the distinguished educationalist, sociologist, and advocate of *yousheng*, who studied under the supervision of Charles Davenport (1866–1944) at Columbia University, returned to China in the 1920s. Pan published many articles and books, and established a journal titled "*Youshengyuekan*" ("Eugenics Monthly") in Shanghai in 1931 to raise public awareness and involve the state in the social reform program (Li 1999; Chung 2010; Sihn 2010). In contemporary China, Pan is respected as the "Father of *yousheng*."

In Pan's time, most Chinese people tended to accept *yousheng* as a way of assisting nation building. As a result, Chinese intellectuals supported the proposition that *yousheng* could benefit the population. On this basis, the Kuomintang government introduced legislation related to contraception, marriage, and sterilization, as well as the isolation of "hereditary defaulters" in 1945 (Li 1999; Chung 2010; Sihn 2010). Some authors explain the limited implementation of *yousheng* practices with poverty, as well as semifeudal and semicolonial structures (Li 1999; Chung 2010; Sihn 2010).

Furthermore, after the new republic China founded in 1949 by the Chinese Communist Party began a comprehensive study following the Soviet Union, both the genetics and the social science communities despised anything tainted with "eugenics" as pseudoscience associated with reactionary bourgeois ideology. For the three following decades, *yousheng*, as well as modern genetics, remained a restricted field (Wu 1983; Wu 1984, 155; Li 1999; Ruan 2002; Chakravarti 2004).

From 1978 on, with emergence of reforms and the "one-child per family" policy, some well-known Chinese medical geneticists and scholars like Wu Min, Lu Hui-Ling, Ruan Fang-Fu, and Li Chong-Gao actively appealed to re-establish *yousheng* studies (Wu 1983; Ruan 2002; Wu 2002). In 1980, an entry on "*youshengkexue*" (eugenics science) re-appeared in the regular publication annals of the *China Encyclopedia Yearbook*; *yousheng* was subsequently involved not only in the conventional fields of sociology, law, ethics, biology, medicine, and genetics but also in related fields such as population studies, neuroscience, obstetrics, pediatrics, psychiatrics, environmental science, and toxicology. *Youshengkexue* in China subsequently evolved into a large interdisciplinary system, classified into four aspects—*jichuyoushengkexue* (basic eugenics science), *youshengkexue* (social eugenics science), *linchuangyoushengkexue* (clinical eugenics science), and *huanjingyoushengkexue* (environmental eugenics science)—to encompass multiple disciplines (Wu 1983; Li 1999; Ruan 2002; Wu 2002; Chung 2014). Li Chong-Gao, a medical geneticist, set up an academic monthly journal named *Chinese Journal of Birth Health and Heredity* (now the official journal of *China Healthy Birth Science Association*) in 1982. In 1988, the *China Healthy Birth Science Association* was formally established, and the former minister of the Department of Health China, Qian Xin-Zhong, was elected as president. Interestingly, a rival group of scholars set up an alternative society in 1989, called the *Chinese Association for Improving Birth Outcomes and Child Development*. So far, the *China Healthy Birth Science Association* consists of more than 20,000 members nationwide, including geneticists, obstetricians, pediatricians, physicians, nurses, maternal and child health care providers, family planning staff, etc.

In 1995, when China's Law on Maternal and Infant Health Care was promulgated, it attracted considerable criticism in the West (Harper 1995; Paul and Spencer 1995; Dickson 1998; Normile 1998; Chen et al. 1999; Guo 1999; Li 2000; Wang 2004). Obviously, the law was liable to be misread by westerners because of a flawed official translation (Normile 1998; Chen et al. 1999; Wang 2004). It is definitely *neither* "China's negative eugenics law" *nor* "a harbinger of one of the darkest episodes of modern history" (Reichman et al. 1996; Canadian College of Medical Geneticists 1997; Chen et al. 1999; Wang 2004). For example, Article 10 of the law warns that marriage between individuals with serious genetic disorders and reproduction by the couple should be conducted under the guidance of their physicians in order to prevent diseases among the offspring; it cites the physician's responsibility to advise the couple to be sterilized. While these measures can be interpreted as institutionalized coercion, it is important to note that any pregnancy termination or sterilization has to be agreed to and signed by the couple concerned first, and the medical procedures have to be free of charge (Article 19). Furthermore, *China Disabled Person's Federation*, organized by the government, and *Chinese Organization for Rare Disorders (CORD)*, a non-profit organization specializing in fields of rare diseases, founded by China-Dolls Center for Rare Disorders in 2013, are organizations to protect the patients' human rights.

The Goal of *Yousheng* in China is to Serve People Who Seek Medical Help

The first goal of medicine is to serve the sick ethically, not to consider the discordant voices or to act in obedience to oppressive governments. We must respect human life, protect those with birth defects and genetic disorders, and support public policy from the standpoint of providing support for these individuals as far as educational opportunities, social respect, and care (Garver and Garver 1991; Li et al. 2013; Paul 2014; Collins and Varmus 2015; Lander 2015). Clinical genetics developed after World War II in response to the concerns of families and clinicians about the risk of genetic disease occurring within a family (Hudson 2011; Collins and Varmus 2015; Lander 2015). Pediatricians were consulted about the risk of recurrence of congenital malformations, or childhood genetic disorders, and neurologists were consulted about the risk of healthy individuals developing the neurodegenerative or neuromuscular disorders that affected other members of their families, while some biologists and physical anthropologists collected information that could help to answer such questions (Clarke 2013; Li et al. 2013). The arrival of genomic technologies that permit low-cost genome-wide analyses shows great potential for the delivery of treatments for the rare hereditary disorders as well as the common diseases of our time. They also permit interventions at the population level, such as preconception or antenatal carrier screening, and non-invasive prenatal diagnostic testing, based on the fetal DNA present in maternal blood from early in pregnancy (Clarke 2013; Li et al. 2013; Paul 2014; Collins and Varmus 2015; Lander 2015). Such developments have the power to make substantial changes in the composition of human populations (Clarke 2013). Of course, the core values of genomic medicine are only to allow the parents or prospective parents or clients to make informed decisions without coercion. Respect for client autonomy is the first priority for the professions.

However, in China, the application of medical genetics to clinical practice, especially in the prevention of pediatric genetic disorders, has still not received adequate attention (Liu and Dai 2016). There has not been any uniform planning of clinical and laboratory genetics developments on a national level, and systematic genetic counseling still seems to be at an early stage of development. Clinical genetics has not received due emphasis, medical genetics is still not recognized as one of the medical specialties, and clinical genetics services are unavailable in the majority of cities or counties. The quality of education regarding modern genetics is also inferior (Huang and Qi 2005; Lo 2008; Li et al. 2013). China has the largest population and is the largest developing country in the world, so birth health is a key issue. According to the authoritative data of 2012 from the China National Health and Family Planning Commission, it is estimated that about 5.6% of Chinese newborns are affected by birth defects every year, i.e., the total number of the affected liveborn infants is 900,000 per year (Qin and Zhu 2013). Therefore, it is very important to emphasize the potential benefits of *yousheng* in China. *Yousheng* is truly a term applied to preventive genetic medicine, though it has been misunderstood by many in the West (Normile 1998; Chen et al. 1999; Li 2000).

In February 2014, the China Food and Drug Administration and the China National Health and Family Planning Commission promulgated a new regulation that immediately prohibited genetic testing—even previously approved services, "including prenatal genetic testing, gene sequencing technique-related products, and cutting-edge products and technologies." The far-ranging ban applies to "all medical technology applications requiring detection equipment, medical diagnostic reagents, and related software; and other products, such as for disease prevention, diagnosis, care, treatment, monitoring, health status evaluation, and prediction of genetic diseases." It was said that the government wanted to eliminate the abuse of genetic testing such as so-called "genetic physical examination" and lack of standards and supervision for the market. This decree astonished almost all professional staff and led to chaos, because a lot of genetic tests or prenatal screening/PGD had to be stopped. Under a storm of protest, the officials eventually had to revise the policy that permits the application of DNA sequencing techniques in the genetic diagnosis, prenatal diagnosis, and PGD in some approved hospitals and biotechnology companies after its approval from August 2014. In the words of Deng Xiao-Ping, the former Chinese leader: "Development is the absolute principle."

In September 2015, China abandoned the one-child policy, allowing all couples to have two children. The move was hailed as a major liberalization of the three-decades-old restriction. Even a lot of advanced age couples were prepared to have another child. Although young and advanced maternal ages are associated with different types of birth defects, it is clear that there is a higher risk for chromosome abnormality in fetuses conceived by women of increased age, typically defined as older than 35 years, and maternal age is a risk factor for congenital heart disease, the leading cause of childhood morbidity and mortality despite dramatic clinical advances, even in the absence of any chromosomal abnormality in the newborn (Reefhuis et al. 2004; Morris et al. 2012; Schulkey et al. 2015). Therefore, many severe challenges await *yousheng* in China.

Conclusion

There still is no "orphan drugs" used to treat rare genetic (orphan) diseases. And so far, gene therapy is a clinical therapeutic tool only used to treat a small number of monogenic disorders worldwide. Predictive genomic testing could play a crucial role in preventing genetic defects, especially in China (Huang and Qi 2005; Li et al. 2013; Collins and Varmus 2015; Lander 2015). The cultural background behind *yousheng* has major implications for genetic practice in China. The Chinese should recover the core meaning of *yousheng*: healthy birth and *voluntary* prevention of birth defects and genetic disorders (Dickson 1998; Chen et al. 1999; Wang 2004; Liu and Dai 2016). The decisions about prenatal DNA diagnosis, *in vitro* fertilization, egg donation, preimplanatory diagnosis, and so forth, should be placed in the hands of prospective parents.[1] Doctors should serve the needs of those prospective parents. The politicians should allow doctors to do so (Tannsjo 1998). Today's technological advances have produced ethical dilemmas that create a pressing need to re-examine moral values and virtues (Joly et al. 2016). We should be careful to consider the potential benefits as well as the possible pitfalls breakthroughs in genetic technology can offer.

Note

1 This is of course not an uncontroversial view. However, it is the authors' view that the harm caused by restrictive regulation exceeds the potential harm caused by social coercion by far.

References

Allen, G.E. 2011. "Eugenics and Modern Biology: Critiques of Eugenics. 1910–1945." *Annals of Human Genetics* 75(3):314–25.

Bashford, A., and P. Levine. 2010. *The Oxford Handbook of the History of Eugenics*. Oxford: Oxford University Press.

Canadian College of Medical Geneticists. 1997. "China's Eugenics Law: Position Statement of the Canadian College of Medical Geneticists." *Journal of Medical Genetics* 34(11):960.

Carter, C.O. 1977. "Monogenic Disorders." *Journal of Medical Genetics* 14(5): 316–20.

Chakravarti, A. 2004. "Ching Chun Li (1912–2003): A Personal Remembrance of a Hero of Genetics." *American Journal of Human Genetics* 74(5):789–92.

Chen, Z., Chen, R.B., Qiu, R.Z., Du, R.F., and W.H.Y. Lo. 1999. "Chinese Geneticists Are far from Eugenics Movement." *American Journal of Human Genetics* 65(4):1199.

China National Committee for Terms in Sciences and Technologies. 2016. "Chinese Terms in Sciences and Technologies." Available at http://www.termonline.cn/index.htm. [Accessed December 10, 2016]

Chung, Y.J. 2010. "Eugenics in China and Hongkong: Nationalism and Colonialism, 1890s-1940s." In *The Oxford Handbook of the History of Eugenics*, edited by A. Bashford and P. Levine, 259–73. Oxford: Oxford University Press.

Chung, Y.J. 2014. "Better Science and Better Race? Social Darwinism and Chinese Eugenics." *Isis* 105(4):793–802.

Clarke, A.J. 2013. "Chapter 30: Ethical and Social Issues in Clinical Genetics." In *Emery and Rimoin's Principles and Practice of Medical Genetics*, 6th ed., edited by D.L. Rimoin, R.E. Pyeritz, and B.R. Korf, 1–40. Oxford: Academic Press.

Collins, F.S., and H. Varmus. 2015. "A New Initiative on Precision Medicine." *The New England Journal of Medicine* 372(9):793–5.

Dai, L., Zhu, J., Liang, J., Wang, Y.P., Wang, H., and M. Mao. 2011. "Birth Defects Surveillance in China." *World Journal of Pediatrics* 7(4):302–10.

Dickson, D. 1998. "Congress Grabs Eugenics Common Ground." *Nature* 394(6695):711.

Dikötter, F. 1992. *The Discourse of Race in Modern China*. Stanford: Stanford University Press.

Ekberg, M. 2007. "The Old Eugenics and the New Genetics Compared." *Social History of Medicine* 20(2):263–80.

Garver, K.L., and B. Garver. 1991. "Eugenics: Past, Present, and the Future." *American Journal of Human Genetics* 49(5):1109–18.

Guo, S.W. 1999. "Cultural Difference and the Eugenics Law." *American Journal of Human Genetics* 65(4):1197–9, author reply 1199–1201.

Harper, P.S. 1995. "Eugenics in China." *Lancet* 346(8973):508.

Ho, D.Y.F. 1985. "Prejudice, Colonialism, and Interethnic Relations: An East-West Dialogue." *Journal of Asian and African Studies* 20(3–4):218–31.

Huang, T.S., and M. Qi. 2005. "Report—21st Century Medical Genetic and Genomic Medicine in China." *Journal of Zhejiang University Science B* 6(12):1223–6.

Hudson, K.L. 2011. "Genomics, Health Care, and Society." *The New England Journal of Medicine* 365(11):1033–41.

Joly, Y., So, D., Saulnier, K., and S.O. Dyke. 2016. "Epigenetics ELSI: Darker Than You Think?" *Trends Genet* 32(10):591–2.

Lander, E.S. 2015. "Cutting the Gordian Helix—Regulating Genomic Testing in the Era of Precision Medicine." *The New England Journal of Medicine* 372(13):1185–6.

Li, C.G. 1980. "Eugenics: Historical Review and Prospectives." *Chinese Journal of Birth Health & Heredity* (1):14–20.

Li, C.G. 1999. "Pan Guang-Dan's Eugenics Thought: For Commemorating the One Hundredth Anniversary of Pan Guang-Dan's Birth." *Chinese Journal of Birth Health & Heredity* 7(6):1–3.

Li, C.C. 2000. "Progressing From Eugenics to Human Genetics. Celebrating the 70th Birthday of Professor Newton E. Morton." *Human Heredity* 50(1):22–33.

Li, P.N., Zhang, H.Z., Li, M.M., Yu, C.L., Qi, M., Yu, J.W., and B.L. Wu. 2013. "Progress and Perspective of Professional Training in Medical Genetics and Genomics: A Report of the Association of Chinese Geneticists in America." *North American Journal of Medical Science* 6(4):173–80.

Liu, W., and Y. Dai. 2016. "'Super Mum' Determined to Cure Children's Illness." *China Daily*, September 9, 2016.

Lo, W.H.Y. 2008. "Prospect of Medical Genetics in China from a Historical Point of View." *Chinese Medical Sciences Journal* 23(2):65–7.

Morris, J.K., Alberman, E., Mutton, D., and P. Jacobs. 2012. "Cytogenetic and Epidemiological Findings in Down Syndrome: England and Wales 1989–2009." *American Journal of Medical Genetics A* 158A(5):1151–7.

Normile, D. 1998. "Geneticists Debate Eugenics and China's Infant Health Law." *Science* 281(5380):1118–19.Paul, D.B. 2014. "What Was Wrong with Eugenics? Conflicting Narratives and Disputed Interpretations." *Science and Education* 23(2):259–71.

Paul, D.B. 2016. "Reflections on the Historiography of American Eugenics: Trends, Fractures, Tensions." *Journal of the History of Biology* April 6. [Epub ahead of print]

Paul, D.B., and H.G. Spencer. 1995. "The Hidden Science of Eugenics." *Nature* 374(6520):302–4.

Qin, H.J., and J. Zhu. 2013. *National Stocktaking Report on Birth Defect Prevention.* Beijing: People's Medical Publishing House.

Reefhuis, J., and M.A. Honein. 2004. "Maternal Age and Non-Chromosomal Birth Defects, Atlanta—1968–2000: Teenager or Thirty-Something, Who Is at Risk?" *Birth Defects Research Part A: Clinical and Molecular Teratology* 70(9):572–9.

Reichman, J.M., Brezis, M., and A. Steinberg. 1996. "China's Eugenics Law on Maternal and Infant Health Care." *Annals of Internal Medicine* 125(5):425–6.

Ruan, F.F. 2002. "A Short History on the Reconstruction of Eugenics in Contemporary China." *China Historical Materials of Science and Technology* 23(4):308–13.

Schulkey, C.E., Regmi, S.D., Magnan, R.A., Danzo, M.T., Luther, H., Hutchinson, A.K., Panzer, A.A., Grady, M.M., Wilson, D. B., and P.Y. Jay. 2015. "The Maternal-Age-Associated Risk of Congenital Heart Disease Is Modifiable." *Nature* 520(7546):230–3.

Sihn, K.H. 2010. "Eugenics Discourse and Racial Improvement in Republican China (1911–1949)." *Uisahak* 19(2):459–85.

Tannsjo, T. 1998. "Compulsory Sterilisation in Sweden." *Bioethics* 12(3):236–49.

Wang, Y.G. 2004. "Chinese 'Eugenics': Definition, Practice and Cultural Values." In *Genomics in Asia: A Clash of Bioethical Interests?* edited by M. Sleeboom, 281–99. London: Kegan Paul.

Wu, M. 1983. "A Talk on Eugenics." *Chinese Journal of Birth Health & Heredity* (1):14–22.

Wu, M. 1984. "Eugenics." In *Chinese Encyclopedia of Medicine (Medical Genetics)*, edited by H.L. Lu, 155–7. Shanghai: Shanghai Science and Technology Press.

Wu, M. 2002. "A Supplement to 'A Short History on the Reconstruction of Eugenics in Contemporary China'." *China Historical Materials of Science and Technology* 23(4):314–15.

Wynbrandt, J., and M. Ludman. 2008. *The Encyclopedia of Genetic Disorders and Birth Defects*, 3rd ed., 150–1. New York: Facts On File.

Part III

Research Challenges

10 Personalized Antidepressant Prescription

A Historical Perspective on Risks and Opportunities[1]

Francesco Spöring

Introduction

Depression is a global concern. The various conditions subsumed under the term "depressive disorders" are estimated to affect more than 350 million people globally (Lépine and Briley 2011; WHO 2012). Among the numerous treatments available for this phenomenon, this chapter is concerned with perhaps the most intensely discussed approach, namely pharmacotherapy. One of the principal problems with this mode of treatment is that not all depressed patients respond equally to the antidepressant drugs available (Horstmann and Binder 2009; Repantis et al. 2009; Gøtzsche 2013). This is where personalized medicine comes into play in promising to provide "the right treatment for the right person at the right time" (Kroemer 2013, 12). To reach this goal, both the prediction of drug efficacy as well as the question of how to identify the "right" person are central.

This chapter explores the development of therapies based on antidepressant drugs and the search for specific clinical profiles designed to predict the likelihood of a desired response to treatment. Due to the possibility that each individual may experience her *Lebenswelt* in a unique way, this search is all the more difficult. In this context, psychiatrists may refer to Karl Jaspers' ([1913] 1973) influential distinction between *explaining* and *understanding*. While explaining refers to causal explanations as used in the natural sciences, understanding denotes the emphatic engagement in a patient's biographically developed uniqueness. In a similar vein, arguing from a political perspective, Hannah Arendt warned of the danger of reducing every human being to a consistent identity comprised of interchangeable reactions—a potential development she described as "total dominance" (Arendt 1962, 438). To what extent is contemporary research on personalized pharmacotherapy reinforcing such a reduction? Together with the uncertain verbal representation of subjective experiences (Derrida 1998), the notion of personal uniqueness raises the question of how individual perceptions and sensations may be described, compared, and quantified. These steps are essential for the task of predicting the efficacy of antidepressant treatment.

There is a plethora of works on depression in the humanities. Prominent sociological explanations for the rising awareness of depression have come from Nikolas Rose (1998, 2003) and Alain Ehrenberg (2013). Running parallel to this tradition, there are critical histories written by former medical practitioners, such as David Healy (1997, 2013) and Peter Gøtzsche (2013). These accounts criticize the influence that pharmaceutical firms appear to exert on clinical trials, and are supported by numerous medical journal articles (see e.g., Ioannidis 2008; Leon 2009, 2010). However, there are also more optimistic views articulated in medical journals with regard to treatment optimization. The promise of a personalized pharmacotherapy implicit in such views could also reinforce a potential epistemic shift in the conception of human individuality. To approach this possibility, I will focus on three interconnected aspects in the development of personalized pharmacotherapy: the *introduction of new antidepressant medicines*, the *development of diagnostic approaches*, and the emerging *measurements of efficacy*. These three facets, which build on each other, will be explored in three parts. Following the established narrative of psychopharmacology, I start with the so-called "therapeutic revolution" of the 1950s. The establishment of operationalized diagnostic criteria and of selective serotonin reuptake inhibitors (SSRIs) during the 1970s and 1980s constitute a second focal point. The last part examines contemporary approaches to predictive antidepressant research. The scope of this historical review does not allow examining all the issues in detail. Among these are:

- The advantages and disadvantages of psychotherapy compared to pharmacotherapy. This is connected to various problems in contemporary trials, such as selective publication (Turner et al. 2008), sponsor bias, a growing placebo effect, and a lack of long-term studies (Ioannidis 2008).
- The distinction between therapy and enhancement and its ethical implications:[2] There is an ongoing debate about the contexts in which the prescription of an antidepressant constitutes therapy, and when it is more appropriate to speak of "cosmetic psychopharmacology" (Kramer 1993; Bjorklund 2005). The latter term denotes the idea that 'healthy' people may be enhanced to become "better than well" (Elliott 2004).

Early Discoveries in the 1950s: A Therapeutic Revolution?

The 1950s are usually depicted as a turning point in psychopharmacology. Not only was the antipsychotic agent chlorpromazine (CPZ) introduced in 1952, it was also the decade in which the first classical antidepressant drugs appeared:[3] In 1956, Roland Kuhn, senior physician at the rural Swiss psychiatric clinic in Münsterlingen, tested several substances as a substitute for CPZ. Among them was the compound G 22355, which had been synthesized eight years previously (Caldwell 1970, 86; Kuhn 2002). To Kuhn's surprise, G 22355 did not produce the antipsychotic effects of CPZ.

However, it eased many symptoms in depressed patients; imipramine hit the market in 1958 (Maxwell 1990). In the same year, iproniazid, an irreversible monoamine oxidase inhibitor (MAOI), was approved for the U.S. market—although only briefly. These new agents marked the beginning of the 'era of antidepressants', since they seemed to have a relatively "stable capacity" (Cartwright and Munro 2010) for producing desired effects in many depressed individuals. In comparison to previous therapies, such as electroshock treatment, it was easier to administer and even seemed to mitigate the suffering of patients who had not been responding to earlier treatments. It is debatable whether the introduction of first generation antidepressants marks a revolution. On the one hand, scholars with a pharmacological background tend to describe the introduction of imipramine and iproniazid as a revolution, leading to improved depression therapies and to new biochemical targets—mainly monoamines (López-Muñoz and Alamo 2009). On the other hand, historians tend to question whether the term "revolution" is appropriate here (Balz 2010). Without any doubt, antidepressants constituted an integral extension of therapeutic options in a multitude of depressed patients.

How did this therapeutic extension affect the concept of psychic individuality? Kuhn, for instance, although impressed by G 22355's general effects, still highlighted the uniqueness of each patient. Consequently, he was reluctant to use numbers as a basis for illustrating the efficacy of imipramine. However, further studies of the drug published in the late 1950s confirmed its efficacy through the quantification of 'successful' cures: Kielholz and Battegay (1958), Lehmann et al. (1958), Fazio et al. (1958), Azima (1959), Ball and Kiloh (1959), Delay et al. (1959), Mann and Macpherson (1959), Sloane et al. (1959), and Straker (1959). The latter provided the first placebo controlled trial with the new drug. Similar quantitative results appeared for iproniazid, which was first marketed as a "psychic energizer" in 1957 (Loomer et al. 1957). While most of these results were based on rather elastic categories, such as "recovered," "vastly improved," or "moderately improved," numeric evaluation instruments in the form of more refined scales won recognition in the 1960s. In 1960, the *Hamilton Rating Scale for Depression (HAMD)* was introduced, and the *Beck Depression Inventory (BDI)* followed a year later. The physician and mathematician Max Hamilton understood his scale as a tool to quickly quantify the results of an interview (Hamilton 1960, 56). While this instrument found, and still finds, numerous applications in pharmacological trials, many psychiatrists have dismissed it due to its reductionist approach (see for instance Fink 2002, 23). Nevertheless, numerous scales of this kind have emerged and consequently reinforced a numeric evaluation of drug effects. These numeric comparisons were enabled by an abstraction of changing behavior patterns to do with sleep, appetite, and activity from individual life stories. Scales such as the HAMD produce scores that reflect the severity of a given depressive state and enable categories to be built at certain cut-offs.

Regarding drug efficacy, scales can be used to illustrate the development of a patient over time, whereas general clinical endpoints such as "response" and "remission" can be defined mathematically, e.g., as a score reduction of at least 50% for a response. Potentially, such approaches devaluate detailed patient accounts of their subjective impressions (Jurk 2005; Tornay 2016).

However, the question of how to determine which patient would truly benefit from which drug still remained. While most studies grouped individual preconditions into two major models, either depressions following unresolved conflicts or depressions as consequences of somatic deficits (Ehrenberg 2013, 139), a variety of other subcategories were created. For instance, Kuhn predicted a particular good effect of imipramin in patients suffering from a "*vital-depressive Verstimmung*" (Kuhn 1958, 449). He vaguely associated this form of depression with observed lethargy, exhaustion, isolation, and "retardation" with regard to action, decision-making, and thinking. In "reactive" and "neurotic" depressions, which seemed to have been brought on by an external event, Kuhn reported a less predictable effect from treatment, particularly when depressions were accompanied by manic states, hallucinations, and epileptic seizures. Kuhn was skeptical about the advent of standardized assessment criteria. In the careful selection of promising subjects, he relied on his experience. As historian Magaly Tornay (2016) points out, in Kuhn's diagnostic approach,[4] listening was more important than seeing. The various expectations underlying his diagnostic process could be criticized as subjective. These potential distortions were to be eliminated by "double-blinded" randomized controlled trials. To Kuhn and many of his German speaking colleagues,[5] the advent of scales as tools for quantifying the severity of seemingly uniform mood disorders meant a loss of specificity in terms of descriptive psychopathology. Despite these reservations, new terms denoting smaller subcategories of depression appeared in the 1960s (Ehrenberg 2013, 127).

In the following decades, with the growing arsenal of antidepressants and increasing experience, recommendations based on diagnostic categories regarding medication of choice and dosage became more refined. At the same time, these drugs' positive effects on many patients led to the hope that they could prevent even more harm if used in prodromal depressions. By the end of the 1950s, prominent physicians such as Frank Ayd (1961) and Paul Kielholz (1965) had published manuals for practitioners to help them 'detect' hidden depressions that could easily be treated with antidepressants. Since such manuals were at least partly supported by pharmaceutical companies, these attempts to expand the diagnoses stirred criticism (e.g., Healy 1997, 162). It was argued that such contributions promoted the broadening of the diagnosis "depressive disorder," as can be seen in the revisions of the *International Statistical Classification of Diseases and Help Related Problems (ICD)* and the *Diagnostic Statistical Manual of Mental Disorders (DSM)*. The first steps toward a standardization of psychopathologic categories had already been initiated through the inclusion of psychiatric phenomena into the *ICD* in 1949, and through the establishment of the *DSM* in 1952 (Tornay 2016, 108).

While following attempts to differentiate subcategories of depression were often related to the choice of 'the right' treatment, psychologist Hans Jürgen Eysenck (1916–1997) proposed an approach to predicting drug effects that were more dimensional than categorical, which means that certain characteristics are described by numeric ranges rather than by categories. Eysenck suggested a number of tests to assess the degree of a person's temperament regarding the dimensions of extraversion and neuroticism (Eysenck 1957, 1963). The respective deviations from a normal value promised to serve as indicators for a clinical spectrum between "dysthymia" and "hysteria." Eysenck's approach promised to provide information about the appropriate drug and dosage. However, as this approach did not take into account 'soft' factors such as set and setting, it has been criticized as too simplistic for clinical practice (Janke 1964). In spite of these shortcomings, subsequent modifications of dimensional temperament assessments, to be discussed in the next section, led to it remaining an influential strand in personalized therapy.

The Neo-Kraepelinian Turn: *DSM-III* and the Advent of SSRIs

By the 1970s, a variety of antidepressants were available. Nine tricyclic antidepressants (TCAs) had been introduced to the market during the 1960s, while MAOIs became unpopular after iproniazid was withdrawn in the U.S. in 1961 due to hepatotoxic side effects. In the early 1970s, the so called "second-generation antidepressants" were introduced. On a theoretical level, the mechanism of action in these substances remained rather unclear. During the 1960s, diagnostic tools such as the spectrophotofluorimetry had reinforced explanations of depression due to a monoaminergic deficiency (López-Muñoz and Alamo 2009). On a clinical level, various psychiatrists described their experience with the antidepressants available in the attempt to determine useful treatment patterns for different patient groups. In their influential book *Psychiatric Syndromes and Drug Treatment*, published in 1979, Nathan S. Kline and Jules Angst recommended various drugs and drug combinations for a number of subcategories of depression.

While the 1960s saw a refinement of application pertaining to dosage and agents of choice, there were particularly heated debates on the validation of psychiatric diagnoses in the 1970s. In the U.S. context, in a landmark article, Eli Robins and Samuel Guze (Robins and Guze 1970) proposed a five-step validation procedure. Guze, Robins, and George Winokur were also involved in the development of the Feighner's criteria. In the context of antipsychiatry, and increasing questioning of the lack of scientific standards in psychiatric diagnoses (Decker 2007; Roelcke 2015), the research group influenced the revision of *DSM-III*, which has often been depicted as the turning point toward neo-Kraepelinianism, following Emil Kraepelin's (1856–1926) attempt to categorize specific symptoms and signs into diagnoses. *DSM-III* introduced explicit operationalized criteria that were more precisely defined than in the previous version. For instance, the term "involutional melancholia," vaguely described in fourteen lines of text in *DSM-II*, was partly replaced by the term "major depressive episode," whose diagnostic criteria were now described on eight pages.[6]

According to Nancy C. Andreasen (2007), it was precisely this *DSM-III* revision that marked a steady decline in descriptive psychopathology (see also Dean 2012; Leon 2013). Later editions of *DSM* seem to have become the authoritative references for validation of diagnostic scales. The subsequent rise in the number of diagnoses of depression is usually interpreted in two ways: On the one hand, it is ascribed to a higher sensitivity regarding identification of depression; on the other hand, it is perceived as an invention of the pharmaceutical industry (Tornay 2016). Regardless of this ambivalence, Andreasen articulated a concern that has been common since the 1970s: the turn toward a "therapeutic empiricism" with its neglect of thorough examinations and of psychoanalytical approaches (Healy 1997, 56).

"Neo-Kraepelinianism" was accompanied by an increasing number of quantitative drug efficacy assessments. Among the numerous psychometric tests and test subvariants developed in the 1970s, the Montgomery–Åsberg Depression Rating Scale (MADRS) stands out, as it began to complement the HAMD from 1978 on. In recent years, most pharmacogenetic studies have employed the latter, while in addition to MADRS, the Clinical Global Impression Scale (GIS) and the Quick Inventory of Depressive Symptomatology (QIDS) are also often applied (Horstmann et al. 2009).[7] Aside from numerically defined diagnostic categories, scales that numeralize certain patients' dimensional characteristics have been consolidated as well. Eysenck's early attempts were followed by various similar tests that are still used to establish relations between individual temperament and dopaminergic, noradrenergic, and serotonergic imbalance.[8] Claude Cloninger's *Tridimensional Personality Questionnaire (TPQ)* (Cloninger et al. 1991; Cloninger 1994), and its successor, the *Temperament and Character Inventory (TCI)*, are still one of the main strains in antidepressant treatment effect prediction (Hruby et al. 2010; Paavonen et al. 2014). U.S. psychiatrist Charles E. Dean (2012) even argues that such dimensional systems are in the process of replacing the classic system of diagnostic categories. These dimensional tests can be complemented with measurements of behavioral characteristics, such as psychomotor behavior, memory-attention, processing speed, and emotional functions to predict response to certain antidepressants (Etkin et al. 2015). Such approaches seem to restrict the importance of verbal patient accounts further.

In the early 1970s, a research team led by David T. Wong (1974) published a landmark study on a new type of antidepressant, which they called *selective serotonin reuptake inhibitor (SSRI)*. Although the pioneering substance, fluoxetine, would become the most famous antidepressant of the 1990s under the name Prozac®, it was not the first *SSRI* marketed: After the short-lived debut of zimelidine in 1982, fluvoxamine followed in 1983. Various federal agencies approved *SSRIs* thereafter: fluoxetine in 1987, citalopram in 1989, sertraline in 1990, and paroxetine in 1991. Compared to tricyclic and atypical antidepressants, *SSRIs* had fewer adverse side effects—in particular, fewer anticholinergic effects—and were easier to titrate due to less interpersonal variability in their pharmacokinetics. Moreover, some psychiatrists also reported that some patients responded better to *SSRIs* than to *TCAs*. Peter D. Kramer's bestseller, *Listening to Prozac* (1993), is a highly visible example. In his collection of case reports, Kramer not only told of patients who had found their "true self" thanks to the drug, he also promoted the prescription of the drug to people whom he had not diagnosed as "depressed." Kramer's introductory case depicted what a profound effect fluoxetine could exert on a patient's well-being. In contrast to this case, a recent meta-analysis of the effects of *SSRIs* on healthy subjects determined only a statistically insignificant increase in well-being (Repantis et al. 2009). Apart from the popular *SSRIs*, other antidepressants such as *serotonin-norepinephrine reuptake inhibitors (SNRIs)*, *norepinephrine reuptake inhibitors (NRIs)*, *norepinephrine-dopamine reuptake inhibitors (NDRIs)*, a *noradrenergic and specific serotonergic antidepressant (NASSA)*, *serotonin-norepinephrine-dopamine reuptake inhibitors (SNDRIs)*, and *multimodal antidepressants* have been developed. All of these classes influence various monoaminergic neurotransmitters, most of them with a more or less comparable general efficacy (Bauer et al. 2007). So, how should physicians choose the optimal drug for a patient? Guidelines such as the World Federation of Societies of Biological Psychiatry Guideline (Bauer et al. 2007) suggest first considering side effect profiles of antidepressants, and then employing various trial and error strategies in case of non-response. However, the molecular genetic methods now available hold the promise of an easier way to find the best treatment for each patient.

A Molecular Revolution? New Biomarkers and Biosignatures

In the past few decades, biochemical explanations for interpersonal variability in drug response have gained broad acceptance. Such theories had already been put forward in the 1950s and 1960s: For instance, Kuhn explained the non-response of some patients by a lack of certain enzymes (Kuhn 1964, 599). Employing such biochemically oriented reasoning, there is burgeoning research on signs in the body and their relevance as a specific biochemical explanation for feelings of malaise. Such signs are often termed "biomarker" or "biosignatures," and suggest objectivity due to a stable reproducibility that does not rely on verbal patient accounts (Wagner 2002).

Among the numerous biomarkers that have been studied as predictors for drug effects, studies focusing on isolated single nucleotide polymorphisms (SNPs)[9] seem to hold much promise for simple recommendations regarding drug selection. For example, some studies advise carriers of a certain allele (5-HTTLPR S-allele) against using *SSRIs* (Porcelli et al. 2012). Other studies focus on the ABCB1 gene, which encodes a protein that is theorized to hinder various antidepressants from staying beyond the *blood brain barrier* (Uhr et al. 2008; Breitenstein et al. 2015). It is speculated that in the case of ABCB1 polymorphisms, non-substrates of P-glycoprotein, such as fluoxetine and mirtazapine, might produce a better response.[10]

Other research focuses on personalized dose adjustment. Most studies in this regard focus on the important drug metabolizing gene CYP2D6, as well as the genes CYP2C19, CYP3A4, and CYP1A2. Specific biochemical tests such as the AmpliChip CYP450® make it possible to group individuals into various categories from "poor metabolizer" to "ultra-rapid metabolizer." This typology may help to prevent overdosing.

Furthermore, hypotheses about a dysregulation of the *hypothalamic-pituitary-adrenal (HPA)* system have become more refined in the last two decades. For instance, a neuropeptide called *Corticotropin-releasing Hormone (CRH)* is theorized to coordinate metabolic adaptation to stress. Thus, a combined dexamethasone-CRH-challenge test could serve as a promising indicator of stress adaptation. Florian Holsboer presumes a uniform, "normal" reaction of a human HPA system to this stress test: If the secretion of the neuropeptides CRH and vasopressin persistently deviates from a "normal HPA set point" (Holsboer 2000, 478), a high likelihood of CRH-dependent depression is assumed. On the one hand, this proposed subcategorization of depression is associated with limited chances of an effective treatment when established agents are concerned. On the other hand, this subcategorization offers a new therapeutic perspective. According to this hypothesis, CRH Receptor Type 1 Agonists or Vasopressin Receptor Agonists should prove to be effective remedies (Ising and Holsboer 2007; Kehne 2007). So far, this approach has not been successful (Schatzberg 2015).

With the exception of minor advances regarding drug choice and dosage as well as the provision of further research directions, the 'molecular revolution' has not yet managed to change clinical practice. However, it affects our language. Over the past six decades, the language used in studies has generally shifted from observant description to molecular references and abstract numbers. This accumulation of acronyms and molecular terms differs considerably from everyday speech. Following the medical discourse requires a familiarity with basic molecular expressions, as well as common psychiatric practices. Due to the time-consuming process of deciphering these terms, it is hard for interested lay persons, as well as many physicians to follow the latest trends—and may get harder in the future. Furthermore, these recent approaches, together with the prognostic advances in neuroimaging (see e.g., van Dinteren et al. 2015), accentuate a rather mechanical concept of human behavior. Quantifiable characteristics outside of a human being's reach seem to be increasingly preferred to patient reports. While this development might be helpful in optimizing drug therapies, at an epistemic level, it comes at the cost of a limited concept of human agency.

Conclusions

From the introduction of imipramine on, a mutual stabilization of drugs, diagnostic criteria, and psychometric assessments can be observed. This process seems to have reinforced an appraisal of numerical quantification and a devaluation of patient individuality. Quantification, while reducing the complexity of reality by neglecting uncountable influences (see e.g., Posternak and Zimmerman 2007), is a precondition of scientific description. However, assessment methods such as *HAMD*, *MADRS*, and *TCI* share a tendency to reduce the complexity of individual life stories to the existence or non-existence of comparatively few characteristics. In biochemically oriented studies on antidepressant efficacy, patient testimonies are typically reduced to fewer than twenty operationalized questions with limited answer options.

As patient reports become less important in the diagnostic process, patients may perceive themselves as passive victims of their biochemical reactions rather than active individuals. To be sure, biochemical imbalances will play an important role in explaining and mitigating certain symptoms. Furthermore, biochemical explanations may relieve patients in terms of stigmatization. However, the still considerable rate of non-responders to antidepressants suggests we would be ill-advised to neglect other possible causes from the psychosocial sphere. Biochemical explanations risk, in Arendt's terms, depicting individuals as homogenous products of interchangeable neuronal reactions. This perception might divert our gaze from other possibilities of actively shaping our individuality, for instance, through a repeated experience of self-efficacy. While such a view may lead to a restricted way of experiencing ourselves as autonomously acting individuals, in many severe forms of depressions, this desired sentiment is even more drastically restricted. In such cases, the principles of beneficence and of non-maleficence may call for 'the right drug for the right patient'. It is therefore essential to pursue operationalized studies based on refined stratification. At a research level, there seems considerable potential to determine criteria for establishing finer patient categories. However, it may be helpful to acknowledge the limits of operational feasibility, and to bear in mind that meaning and significance are co-produced in interpersonal encounters. This is where a stratified medicine becomes personal.

Notes

1 The author would like to thank Mariacarla Gadebusch Bondio, Maria Marloth, and Karl Hughes for their helpful comments.

2 Among the voluminous contributions on ethical implications, it is worth referring to Parens (1998), Elliot (1999), and The President's Council on Bioethics (2003).

3 Before the 1950s, there were many other treatments given, including insulin shock therapy, chloral hydrate, and stimulants, such as cocaine and amphetamines. For an overview see Linde and Defren (1988).
4 Kuhn stood in the tradition of Ludwig Binswanger's *Daseinsanalyse*. The notion of *Dasein* (human existence) was a reference to philosopher Martin Heidegger, from which Binswanger drew heavily (Binswanger 1949).
5 Many German psychiatrists shared this distrust of quantification (Healy 1997, 56; Schwab 2002).
6 However, the term "melancholia" still figured as a vaguely defined epiphenomenon in *DSM-III*.
7 These tests are often validated in comparison to more detailed standard surveys such as the *Composite International Diagnostic Interview (CIDI)* and the *Structured Clinical Interview for DSM-5* (SCID-5). Due to their compatibility with the operationalized criteria listed in both *DSM-5* and *ICD-10*, these interviews are often considered to be the "gold standard" (First 2015).
8 Dopamine, serotonin, noradrenaline, and Gamma-aminobutyric acid are neurotransmitters that play a crucial role in affect regulation.
9 An SNP denotes a deviance of a single nucleotide in reference to a normed DNA-sequence within a population that is determined in several cases. For an overview on current biomarker-driven research directions, see e.g., Labermaier et al. 2013.
10 Various SNPs such as rs2032583, rs2235015, rs2235040, rs1045642, rs2032582, rs112850, C3435T, C1236, and G2677T are associated with the overcoming of the BBB.

References

Andreasen, N.C. 2007. "DSM and the Death of Phenomenology in America: An Example of Unintended Consequences." *Schizophrenia Bulletin* 33(1):108–12.

Arendt, H. 1962. *The Origins of Totaliarism*. Cleveland, New York: Meridian Books.

Ayd, F. 1961. *Recognizing the Depressed Patient*. New York: Grune and Stratton.

Azima, H. 1959. "Imipramine (Tofranil): A New Drug for the Depressed." *Canadian Medical Association Journal* 80(7):535–40.

Ball, J.R., and L.G. Kiloh. 1959. "A Controlled Trial of Imipramine in Treatment of Depressive States." *British Medical Journal* 2(5159):1052–5.

Balz, V. 2010. *Zwischen Wirkung und Erfahrung—eine Geschichte der Psychopharmaka: Neuroleptika in der Bundesrepublik Deutschland; 1950–1980. Science Studies*. Bielefeld: Transcript.

Bauer, M., Bschor T., Pfennig A., Whybrow, P.C., Angst, J., Versiani, M., and H.-J. Möller. 2007. "World Federation of Societies of Biological Psychiatry (WFSBP) Guidelines for Biological Treatment of Unipolar Depressive Disorders in Primary Care." *The World Journal of Biological Psychiatry* 8(2):67–104.

Binswanger, L. 1949. "Die Bedeutung der Daseinsanalytik Martin Heideggers für das Selbstverständnis der Psychiatrie." In *Martin Heideggers Einfluss auf die Wissenschaften: aus Anlass seines sechzigsten Geburtstages*, edited by Carlos Astrada, 58–72. Bern: Francke.

Bjorklund, P. 2005. "Can there Be a 'Cosmetic' Psychopharmacology? Prozac Unplugged: The Search for an Ontologically Distinct Cosmetic Psychopharmacology." *Nursing Philosophy* 6(2):131–43.

Bonah, C., ed. 2009. *Harmonizing Drugs: Standards in 20th-Century Pharmaceutical History*. Paris: Ed. Glyphe.

Breitenstein, B., Brückl, T. M., Ising, M., Müller-Myhsok, B., Holsboer, F., and D. Czamara. 2015. "ABCB1 Gene Variants and Antidepressant Treatment Outcome: A Meta-Analysis." *American Journal of Medical Genetics: Part B, Neuropsychiatric Genetics* 168B(4): 274–83.

Caldwell, A. E. 1970. *Origins of Psychopharmacology: From CPZ to LSD. American Lecture Series 777*. Springfield: Thomas.

Cartwright, N., and E. Munro. 2010. "The Limitations of Randomized Controlled Trials in Predicting Effectiveness." *Journal of Evaluation in Clinical Practice* 16(2):260–6.

Cloninger, C. R. 1994. *The Temperament and Character Inventory (TCI): A Guide to Its Development and Use*, 1st ed. St. Louis: Center for Psychobiology of Personality, Washington University.

Cloninger, C. R., Przybeck, T. R., and D. M. Svrakic. 1991. "The Tridimensional Personality Questionnaire: U.S. Normative Data." *Psychological Reports* 69(3 Pt 1): 1047–57.

Dean, C. E. 2012. "The Death of Specificity in Psychiatry: Cheers or Tears?" *Perspectives in Biology and Medicine* 55(3):443–60.

Decker, H. S. 2007. "How Kraepelinian Was Kraepelin? How Kraepelinian Are the Neo-Kraepelinians? From Emil Kraepelin to DSM-III." *History of Psychiatry* 18(71 Pt 3):337–60.

Delay, J., Deniker, P., Lempreriere, T., Ropert, M., Colin, W., and B. Ogrizek. 1959. "Etude de l'Efficacité de l'Imipramine (G. 22.355) dans le Traitement des États Dépressifs." [Study of the Effectiveness of Imipramine in the Treatment of Depressive States.] *Annales Médico-Psychologiques* 117(3):521–31; discussion 531–5.

Derrida, J., and P. Engelmann, eds. 1998. *Auslassungspunkte: Gespräche. Passagen Philosophie*. Wien: Passagen.

Ehrenberg, A. 2013. *Das erschöpfte Selbst: Depression und Gesellschaft in der Gegenwart*. Frankfurt am Main: Suhrkamp.

Elliott, C. 1999. *A Philosophical Disease: Bioethics, Culture, and Identity: Reflective Bioethics*. New York: Routledge.

Elliott, C. 2004. *Better than Well: American Medicine Meets the American Dream*. New York: Norton.

Etkin, A., Patenaude, B., Song, Y.J.C., Usherwood, T., Rekshan, W., Schatzberg, A. F., Rush, A. J., and L. M. Williams. 2015. "A Cognitive-Emotional Biomarker for Predicting Remission with Antidepressant Medications: A Report from the iSPOT-D Trial." *Neuropsychopharmacology* 40(6):1332–42.

Eysenck, H. J. 1957. "Drugs and Personality. I. Theory and Methodology." *Journal of Mental Science* 103:119–31.

Eysenck, H. J., ed. 1963. *Experiments with Drugs*. Oxford: Pergamon Press.

Fazio, C., Gilberti, F., and C. Loeb. 1958. "Un nuovo trattamento farmacologico (derivato iminodibenzilico G. 22355) nella cura degli stati depressivi." [A New Drug (Iminodibenzylic Derivative, G 22355) in Therapy of Depressive States.] *Minerva Medica* 49(65–6):3143–8.

Fink, M. 2002. "Clinical Evaluation of Psychoactive Drugs in the 21st Century." In *From Psychopharmacology to Neuropsychopharmacology & The Story of the CINP*, edited by T. Ban, D. Healy, and E. Shorter, 21–6. Budapest: Animula.

First, M. B. 2015. "Structured Clinical Interview for the DSM (SCID)." In *The Encyclopedia of Clinical Psychology*, edited by Robin L. Cautin and Scott O. Lilienfeld, 1–6. Chichester, West Sussex: John Wiley and Sons.
Gøtzsche, P. C. 2013. *Deadly Medicines and Organised Crime: How Big Pharma Has Corrupted Healthcare*. London: Radcliffe.
Hamilton, M. 1960. "A Rating Scale for Depression." *Journal of Neurology, Neurosurgery, and Psychiatry* 23:56–62.
Healy, D. 1997. *The Antidepressant Era*. Cambridge. Harvard University Press.
Healy, D. 2013. *Pharmageddon*. Berkeley: University of California Press.
Holsboer, F. 2000. "The Corticosteroid Receptor Hypothesis of Depression." *Neuropsychopharmacology* 23(5): 477–501.
Horstmann, S., and E. B. Binder. 2009. "Pharmacogenomics of Antidepressant Drugs." *Pharmacology & Therapeutics* 124(1):57–73.
Hruby, R., Nosalova, G., and S. Hruba. 2010. "Predictive Significance of TCI-R for Antidepressant Treatment." *Medical Science Monitor* 16(8):CR383–8.
Insel, T., Cuthbert, B., Garvey, M., Heinssen, R., Pine, D. S., Quinn, K., Sanislow, C., and P. Wang. 2010. "Research Domain Criteria (RDoC): Toward a New Classification Framework for Research on Mental Disorders." *The American Journal of Psychiatry* 167(7):748–51.
Ioannidis, J.P.A. 2008. "Effectiveness of Antidepressants: An Evidence Myth Constructed from a Thousand Randomized Trials?" *Philosophy, Ethics, and Humanities in Medicine: PEHM* 3:14.
Ising, M., and F. Holsboer. 2007. "CRH-sub-1 Receptor Antagonists for the Treatment of Depression and Anxiety." *Experimental and Clinical Psychopharmacology* 15(6):519–28.
Janke, W. 1964. *Experimentelle Untersuchungen zur Abhängigkeit der Wirkung psychotroper Substanzen von Persönlichkeitsmerkmalen: Ein Beitrag zur Begründung einer differentiellen Pharmakopsychologie*. Frankfurt am Main: Akad. Verl.-Ges.
Jaspers, K. 1973. *Allgemeine Psychopathologie*. Berlin: Springer.
Jurk, C. 2005. "Der niedergeschlagene Mensch: Depression. Eine sozialwissenschaftliche Studie zu Geschichte und gesellschaftlicher Bedeutung einer Diagnose." Dissertation, Justus-Liebig-Universität Gießen.
Kehne, J. H. 2007. "The CRF1 Receptor, a Novel Target for the Treatment of Depression, Anxiety, and Stress-related Disorders." *CNS & Neurological Disorders Drug Targets* 6(3): 163–82.
Kielholz, P. 1965. *Diagnose und Therapie der Depressionen für den Praktiker*. München: J. F. Lehmanns.
Kielholz, P., and R. Battegay. 1958. "Behandlung depressiver Zustandsbilder, Unterspezieller Berücksichtigung von Tofranil, einem neuen Antidepressivum." [Treatment of Depressive States with Special Consideration of Tofranil, a New Antidepressant.] *Schweizerische medizinische Wochenschrift* 88(31):763–7.
Kline, N. S., and J. Angst. 1979. *Psychiatric Syndromes and Drug Treatment*. New York: Aronson.
Kramer, P. D. 1993. *Listening to Prozac*. New York: Viking.
Kroemer, H. K. 2013. "Personalisierte Medizin—zum Stand der Forschung." In *Deutscher Ethikrat: Personalisierte Medizin—Der Patient als Nutznießer oder als Opfer?* 11–22. Available at http://www.ethikrat.org/dateien/pdf/tagungsdokumentation-personalisierte-medizin.pdf. [accessed November 26, 2015].

Kuhn, R. 1958. "The Treatment of Depressive States with G 22355 (Imipramine Hydrochloride)." *The American Journal of Psychiatry* 115(5):459–64.

Kuhn, R. 1964. "5 Jahre medikamentöser Behandlung depressiver Zustände mit Iminodibenzylderivaten." *Schweizerische medizinische Wochenschrift* (94): 590–601.

Kuhn, R. 2002. "Experiments with G 223355, Tested on Approx. 150 Patients: Correspondence with Robert Domenjoz, August 11, 1956." In *From Psychopharmacology to Neuropsychopharmacology & the Story of the CINP*, edited by T. Ban, D. Healy, and E. Shorter, 317–19. Budapest: Animula.

Labermaier, C., Masana, M., and M.B. Müller. 2013. "Biomarkers Predicting Antidepressant Treatment Response: How Can We Advance the Field?" *Disease Markers* 35(1):23–31.

Lehmann, H.E., Cahn, C.H., and R.L.d. Verteuil. 1958. "The Treatment of Depressive Conditions with Imipramine (G 22355)." *Canadian Psychiatric Association Journal* 3(4):155–64.

Leon, J. d. 2009. "The Future (or Lack of Future) of Personalized Prescription in Psychiatry." *Pharmacological Research* 59(2):81–9.

Leon, J. d. 2010. "Future of Personalized Prescription in Psychiatry." In *Pharmacogenomics in Psychiatry*, Vol. 25, edited by Matthias Schwab, 118–34. Advances in Biological Psychiatry 25. Basel: KARGER.

Leon, J. d. 2013. "Is Psychiatry Scientific? A Letter to a 21st Century Psychiatry Resident." *Psychiatry Investigation* 10(3):205–17.

Lépine, J.-P., and M. Briley. 2011. "The Increasing Burden of Depression." *Neuropsychiatric Disease and Treatment* 7(Suppl. 1):3–7.

Linde, Otfried K., and Günter Defren, eds. 1988. *Pharmakopsychiatrie im Wandel der Zeit: Erlebnisse u. Ergebnisse.* Klingenmünster: Tilia-Verl. Mensch u. Medizin.

Loomer, H.P., Saunders, J.C., and N. S. Kline. 1957. "A Clinical and Pharmacodynamic Evaluation of Iproniazid as a Psychic Energizer." *Psychiatric Research Reports* 8:129–41.

López-Muñoz, F., and C. Alamo. 2009. "Monoaminergic Neurotransmission: The History of the Discovery of Antidepressants from 1950s until Today." *Current Pharmaceutical Design* 15(14):1563–86.

Mann, A.M., and A.S. Macpherson. 1959. "Clinical Experience with Imipramine (G22355) in the Treatment of Depression." *Canadian Psychiatric Association Journal* 4(1):38–47.

Maxwell, R.A., and S.B. Eckhardt. 1990. *Drug Discovery: A Casebook and Analysis*. Clifton: Humana Press.

Paavonen, V., Kampman, O., Illi, A., Viikki, M., Setälä-Soikkeli, E., and E. Leinonen. 2014. "A Cluster Model of Temperament as an Indicator of Antidepressant Response and Symptom Severity in Major Depression." *Psychiatry Investigation* 11(1):18–23.

Parens, Erik, ed. 1998. *Enhancing Human Traits: Ethical and Social Implications. Hastings Center Studies in Ethics*. Washington: Georgetown University Press.

Porcelli, S., Fabbri, C., and A. Serretti. 2012. "Meta-Analysis of Serotonin Transporter Gene Promoter Polymorphism (5-HTTLPR) Association with Antidepressant Efficacy." *European Neuropsychopharmacology* 22(4):239–58.

Posternak, M.A., and M. Zimmerman. 2007. "Therapeutic Effect of Follow-Up Assessments on Antidepressant and Placebo Response Rates in Antidepressant Efficacy Trials: Meta-Analysis." *The British Journal of Psychiatry* 190:287–92.

The President's Council on Bioethics, ed. 2003. *Beyond Therapy: Biotechnology and the Pursuit of Happiness*. Washington: Regan Books.
Repantis, D., Schlattmann, P., Laisney, O., and I. Heuser. 2009. "Antidepressants for Neuroenhancement in Healthy Individuals: A Systematic Review." *Poiesis Prax* 6(3–4):139–74.
Robins, E., and S.B. Guze. 1970. "Establishment of Diagnostic Validity in Psychiatric Illness: Its Application to Schizophrenia." *The American Journal of Psychiatry* 126(7):983–7.
Roelcke, V. 2015. "Forschungsintrument und Normierungsinstanz: Zur Ambivalenz psychiatrischer Klassifikationen." In *Norm als Zwang, Pflicht und Traum: Normierende versus individualisierende Bestrebungen in der Medizin*, edited by E. Brinkschulte and M. Gadebusch Bondio, 155–81. Frankfurt am Main: PL Academic Research.
Rose, N. 1998. *Inventing Our Selves: Psychology, Power and Personhood: Cambridge Studies in the History of Psychology*. Cambridge: Cambridge University Press.
Rose, N. 2003. "Neurochemical Selves." *Society* 41(1):46–59.
Schatzberg, A.F. 2015. "Issues Encountered in Recent Attempts to Develop Novel Antidepressant Agents." *Annals of the New York Academy of Sciences* 1345: 67–73.
Schwab, J.J. 2002. "Psychopharmacology 1980: Retrospect and Prospect." In *From Psychopharmacology to Neuropsychopharmacology in the 1980s and the Story of CINP, as Told in Autobiography*, edited by T.A. Ban, Healy, D., and E. Shorter, 49–64. History of Psychopharmacology and the CINP, as Told in Autobiography, Vol. 3. Budapest, Hungary: Animula Pub. House.
Siegle, G.J., Thompson, W.K., Collier, A., Berman, S.R., Feldmiller, J., Thase, M.E., and E.S. Friedman. 2012. "Toward Clinically Useful Neuroimaging in Depression Treatment: Prognostic Utility of Subgenual Cingulate Activity for Determining Depression Outcome in Cognitive Therapy across Studies, Scanners, and Patient Characteristics." *Archives of General Psychiatry* 69(9):913–24.
Sloane, R.B., Habib, A., and U.E. Batt. 1959. "The Use of Imipramine (Tofranil) for Depressive States in Open Ward Settings of a General Hospital: A Preliminary Report." *Canadian Medical Association Journal* 80(7):540–6.
Straker, M. 1959. "Imipramine (Tofranil): A Safe, Effective Antidepressant Drug in Private Practice." *Canadian Medical Association Journal* 80(7):546–9.
Tornay, M. 2016. *Zugriffe auf das Ich: Psychoaktive Stoffe und Personenkonzepte in der Schweiz, 1945 bis 1980*. S. l.: Eisenbrauns.
Turner, E.H., Matthews, A.M., Linardatos, E., Tell, R.A., and R. Rosenthal. 2008. "Selective Publication of Antidepressant Trials and Its Influence on Apparent Efficacy." *The New England Journal of Medicine* 358(3):252–60.
Uhr, M., Tontsch, A., Namendorf, C., Ripke, S., Lucae, S., Ising, M., Dose T., Ebinger, M., Rosenhagen, M., Kohli, M., Kloiber, S., Salyakina, D., Bettecken, T., Specht, M., Pütz, B., Binder, E. B., Müller-Myhsok, B., Holsboer, F. 2008. "Polymorphisms in the Drug Transporter Gene ABCB1 Predict Antidepressant Treatment Response in Depression." *Neuron* 57(2):203–9.
van Dinteren, R., Arns, M., Kenemans, L., Jongsma, M.L.A., Kessels, R.P.C., Fitzgerald, P., Fallahpour, K., Debattista, C., Gordon, E., and L.M. Williams.

2015. "Utility of Event-Related Potentials in Predicting Antidepressant Treatment Response: An iSPOT-D Report." *European Neuropsychopharmacology* 25(11):1981–90.
Wagner, J. A. 2002. "Overview of Biomarkers and Surrogate Endpoints in Drug Development." *Disease Markers* 18(2):41–6.
WHO. 2012. "Fact Sheet N°369." Available at http://www.who.int/mediacentre/factsheets/fs369/en/ [accessed August 27, 2015].
Wong, D. T., Horng, J. S., Bymaster, F. P., Hauser, K. L., and B. B. Molloy. 1974. "A Selective Inhibitor of Serotonin Uptake: Lilly 110140, 3-(p-trifluoromethylphenoxy)-N-methyl-3-phenylpropylamine." *Life Sciences* 15(3):471–9.

11 Predicting, Preventing, and Treating Alzheimer's Disease

Current State and Future Challenges

Stefan F. Lichtenthaler

Basic Facts about Alzheimer's Disease

According to the *Guinness Book of Records*, the French woman Jeanne Calment is the oldest person to have lived in the world. In 1997 she died at the age of 122 years. However, the mean life expectancy in the Western world is around 80 years. Why do we not all reach Jeanne Calment's age? This is due to the various diseases which are the main causes of mortality in older age in industrialized countries. The top ten diseases include cancer and cardiovascular diseases, but also Alzheimer's disease (AD). AD is the most common form of dementia, and affects over twenty million patients worldwide (Ballard et al. 2011). It is characterized by progressive loss of memory, declining cognitive function, and, ultimately, it leads to decreasing physical functions and death. The life expectancy after AD diagnosis is typically between four and seven years, but can vary significantly in individual cases. In the last years of life an AD patient is bed-ridden, and requires constant caregiver support, making AD not only a major burden for the patient and the families, but also for our health care systems.

AD is an age-associated disease. At the age of 65 relatively few patients are affected by AD. However, above 65 the numbers of AD cases increase almost exponentially with increasing age. In the age group 85 and older, more than 40% of the population suffers from AD (Hebert et al. 2003). Thus, increasing age is the major risk factor for AD. This is explained in more detail below, where the molecular causes of the disease will be described. Because life expectancy is increasing, and the baby-boomers are reaching the AD-relevant age, the numbers of AD patients are expected to double by 2050 (Hebert et al. 2003). Additional AD risk factors, besides increasing age, are environmental risk factors, as well as the genetic risk factors that are described below.

Molecular Causes of the Disease and Risk Factors

Over 100 years ago—in 1907—the German psychiatrist Alois Alzheimer described the first case of a new dementia now known as AD (Alzheimer

1907). Upon autopsy of his first AD patient, he observed an extensive loss of nerve cells as well as two forms of protein deposits in the brain, the amyloid plaques and the neurofibrillary tangles. Today, plaques and tangles are still used for the definite diagnosis of AD upon autopsy, and are major neuropathological hallmarks of the disease. The early molecular steps preceding plaque and tangle formation are assumed to cause the disease (Figure 11.1), although additional possibilities are also discussed (Herrup 2010; Huang and Mucke 2012). The whole process, referred to as the amyloid cascade (Hardy and Selkoe 2002), is widely accepted by most scientists studying AD, and starts with a protein called the amyloid precursor protein (APP). APP is found in the membrane of the cells in our body. Two enzymes, called β- and γ-secretase (Fluhrer et al. 2009; Vassar et al. 2014), act as molecular scissors and cleave APP at two sites, resulting in the release of a small peptide, called the amyloid β peptide (Aβ, depicted as red cylinders in Figure 11.1). Due to its biophysical properties, Aβ has the tendency to aggregate. Small aggregates, called Aβ-oligomers, are neurotoxic, and start damaging the nerve cells and their contact sites (synapses) in the brain (Haass and Selkoe 2007). This induces aggregation of a second protein, called tau, and is followed by inflammatory processes in the brain, finally resulting in the death of the nerve cells and the onset of AD symptoms. During this process the Aβ aggregates continue to grow and are finally visible upon autopsy as the characteristic plaques. Likewise, the tau aggregates grow, leading to the appearance of the second neuropathological AD hallmark, the neurofibrillary tangles (Figure 11.1). Aggregate formation starts in specific brain areas, such as the locus coeruleus or the entorhinal cortex in the case of tau, and then spreads to other brain regions as well. At the onset of the disease symptoms, over 50% of the nerve cells are already lost in a specific brain area, the layer II of the entorhinal cortex (Gomez-Isla et al. 1996). This demonstrates that the destructive process in the brain starts long (presumably twenty years) before the onset of the symptoms. Thus, if we were able to predict who gets AD ten or twenty years later, we should be able to prevent AD, or at least slow down its progression, provided that suitable drugs are available. The prediction of AD will be discussed in more detail bellow. The situation is similar to high blood pressure or high blood cholesterol levels, which are major risk factors for cardiovascular disease. Drugs to reduce both symptoms are among the most widely sold drugs.

Another point to consider is the following: Nerve cells, which die in the brain during the course of AD, can presently not be replaced. As a result, a treatment that starts after onset of AD symptoms may be able to slow down, or even block the further destruction in the brain, but is unlikely to be able to fully restore brain functions. Therefore, as with cardiovascular diseases, prevention as well as treatment is a central concern in AD research.

Importantly, formation of the Aβ peptide (red cylinders in Figure 11.1) is a normal, physiological process, and happens in everybody's brain (Haass and Selkoe 2007). Our brain can remove and degrade Aβ. However, as we

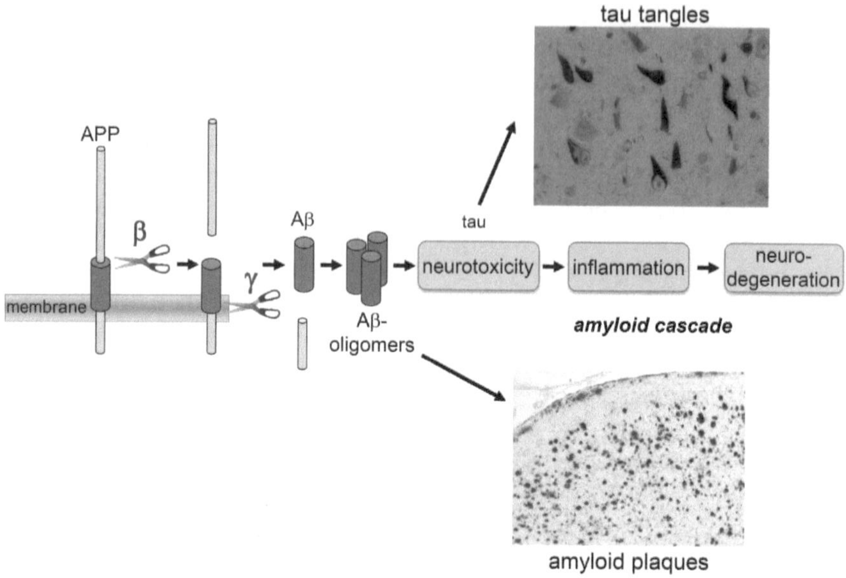

Figure 11.1 For details see text. APP: amyloid precursor protein; β, γ: β-secretase and γ-secretase.

age, Aβ removal appears to be less efficient in our brain, similar to other body processes which also slow down with increasing age. As a result, brain Aβ levels increase, and plaque formation slowly starts. Plaques and tangles are found in most elderly brains. This can be seen, for example, when an elderly person dies in an accident—but not of AD—and the brain is subsequently analyzed after autopsy. In contrast to non-AD brains, a typical AD brain is full of plaques and tangles. New brain imaging methods, such as positron emission tomography, even allow us to see the plaques in living patients. This method can be used several years before the onset of the disease to identify individuals at high risk of developing the disease symptoms a few years later. The finding of low numbers of plaques and tangles, even in non-AD brains, demonstrates that in principle anybody may get AD if they live long enough.

Environmental factors, which are not yet well understood, along with genetic factors appear to increase the speed of the amyloid cascade such that affected individuals will develop the disease during their lifetime, while others die from other diseases before reaching the age at which AD would manifest in their brain. One genetic risk factor speeding up the disease process is apolipoprotein E. The gene exists in three slightly different forms, referred to as ε2, ε3, and ε4. Individuals with the ε4 variant have an approximately four to eight times increased risk of AD (Kanekiyo et al. 2014). This particular form of apolipoprotein E is found in about half of all AD patients.

Other genetic risk factors have also been identified, but are not yet as well understood as apolipoprotein E (Holtzman et al. 2011; Karch et al. 2014). Not everyone who carries one or several risk factors will definitely get the disease. This is similar to smoking, which is a risk factor for lung cancer. But some people, such as the heavy smokers Winston Churchill and Helmut Schmidt, did not get lung cancer. In contrast to genetic risk factors, there are disease-causing mutations, such as in the gene encoding APP or the molecular scissor γ-secretase. An individual carrying such a mutation will develop the disease if they live long enough. As a result of the disease-causing mutations, the aggregating Aβ is generated faster. Thus, AD starts earlier in life, which can be as early as 18, but frequently starts after 50. This inherited form of AD accounts for less than 1% of all AD cases, but the mutations can be used to definitely predict who will develop AD later. Up to now, over 100 different mutations have been identified in the APP and the γ-secretase gene. Within one family, all affected members carry the same mutation, and may show the first symptoms at a similar age, thus allowing a relatively precise prediction of the age of onset. The families affected by disease-causing mutations are essential for the preventive clinical trials for AD, which will be described in more detail below.

The good news about genetics is that protective genetic mechanisms exist. A small proportion of the population in Iceland has a mutation in the APP gene which makes it less susceptible to cleavage by β-secretase, resulting in lower Aβ production (Jonsson et al. 2012). Whether Jeanne Calment carried this protective mutation or another with similar effects, is unknown. But it is possible that additional protective mutations are present in centenarians.

Current and Future Treatment Options

Currently, it is not yet possible to treat the causes of AD, such as preventing Aβ and tau aggregation or neuroinflammation. Upon AD diagnosis, a patient is treated for the symptoms, in particular memory deficit. The first line of drugs blocks an enzyme called acetyl-cholinesterase, which neutralizes a specific neurotransmitter in the brain and ensures that the transmission of information between nerve cells happens normally (Anand and Singh 2013). In AD patients there is too little of this neurotransmitter, resulting in reduced information transmission in the brain. Thus, drugs blocking acetylcholinesterase should increase the transmission of information in the brain, and are widely used to alleviate the memory loss. Yet, we need to be aware that these drugs are not able to block the amyloid cascade, or to prevent further nerve cell loss or even death.

Intensive research efforts are under way to improve the treatment options for AD. Given that the molecular mechanisms leading to AD start about twenty years before the onset of the disease, one of the goals of current research is to block the process early enough to prevent AD, or at least to slow down the speed of its development significantly so that disease onset can be postponed. All stages in the amyloid cascade (Figure 11.1) are targeted with drug development programs in academia and in pharmaceutical companies. The most advanced approaches are blockers of one of the molecular scissors, the β-secretase, and a possible vaccination against AD. While β-secretase blockers reduce Aβ generation, the vaccine reduces Aβ, and in particular its neurotoxic aggregates after they have been made (Schenk et al. 2012; Vassar et al. 2014). Like drug development programs for other diseases, both approaches have suffered some setbacks, but are now far advanced in phase III clinical trials. This raises the hope that within a few years the first drugs targeting the causes and not just the symptoms of AD will be available. However, in order to be useful as preventive drugs, we need to have new diagnostic ways of predicting who will suffer AD a few years later.

Prediction of AD

Individuals at high risk of cardiovascular disease can be relatively easily identified, for example with measurement of blood pressure or blood cholesterol levels. Such easy measurements are not available for AD, where the diagnosis—after the onset of symptoms—is made with a variety of medical exams, including psychological tests and measurement of Aβ and tau levels in cerebrospinal fluid. But how do we predict who will get AD, and consequently should take drugs targeting the causes of AD while remaining presymptomatic? While such diagnostics are not yet routinely available, they are currently being developed. The most promising methods for a diagnostic prediction of AD are measurement of biomarkers in body fluids, in particular in cerebrospinal fluid (as described in detail in another chapter in this book), and brain imaging (Sutphen et al. 2014; Hendrix et al. 2015). Brain imaging uses radioactive tracers, which tell us to what extent the brain is covered with plaques or tangles. Amyloid plaque imaging is able to predict relatively accurately who is at high risk of developing AD ten years later. However, this method is still very expensive, and only available at highly specialized centers, such as university hospitals.

What are benefits and drawbacks of an AD prediction? The drawback is an ethical concern. If an individual receives the information that they are likely to develop AD ten years later, this information can be devastating, in particular if no preventive drug can as yet be offered. But this may change as preventive drugs become available. In contrast, the advantages of a reasonable prediction are manifold. It may allow 'future patients' (or 'virtual sick') to plan their lives accordingly, for example in personal and legal terms. The 'future patient' may take non-drug-based preventive measures, such as increased physical exercise, reducing cardiovascular risk factors, or a healthy diet. Additionally, the individual may determine who should be their legal representative once he or she is no longer able to carry out basic processes in life. Yet another point—relevant for the inherited AD cases—is in vitro fertilization and preimplantation diagnostics, where the inheritor is still in reproductive age, and considers having a child without the AD-causing mutation. It is clear that these topics are ethically controversial, and thus not easy to resolve.

Another advantage of AD prediction is for preventive clinical trials testing new drugs targeting the causes of AD. If we do not know who will get the disease five to ten years later, many thousands of individuals would need to be included in clinical studies, and then be tracked for many years, which is financially and logistically nearly impossible. Including only high risk individuals in the clinical trials has the added value that individuals with low risk would not be included, and thus not be exposed to potential side effects of the drugs tested in the trials. Thus, brain imaging can be used to predict AD, and to include only individuals at high risk in the preventive studies.

Another group of future AD patients are those carrying a gene mutation causing inherited AD. In these cases, it is not only clear that they will get the disease, but frequently—based on their family history—a prediction can be made regarding the age at which they will get the symptoms. Such preventive trials have recently started. One of them is DIAN—the dominantly inherited Alzheimer network, which has recruited carriers of diverse APP and γ-secretase mutations in different countries (Moulder et al. 2013). Another consortium is a Colombian study, which focuses on a large family in Colombia carrying a specific γ-secretase mutation (Reiman et al. 2011). Participants in these studies started treatment several years before the expected age of onset of the disease symptoms. If the drugs are effective, they will delay the onset of the disease symptoms; we should know in a few years' time.

Not all drugs work equally well on all patients. Consequently, some patients may only experience the side effects of a drug while not enjoying the beneficial effects of the drug. Thus, another aspect of prediction is the prognosis regarding which patient will benefit from a drug. Let us take the example of β-secretase blockers. If we apply diagnostic tests (companion diagnostics) that quickly tell us whether the drug lowers Aβ levels in an individual, then this individual is likely to experience the beneficial effects, i.e., a slowing down of the AD pathogenesis. One such test is to measure Aβ levels in CSF, which has been used for many years in clinical trials. Additional tests are currently being developed, for example, targeting the efficacy of the β-secretase enzyme.

Conclusion and Outlook

AD is the most common form of dementia, and treating AD is a major challenge for the 21st century. While basic and applied research are yielding a steadily clearer picture of the molecular causes of the disease, and offer new drug targets, drugs are being developed that target the causes of AD, and are currently being tested in first prevention trials. Progress is also being made in developing novel diagnostics for AD, which is particularly important in predicting 'future AD patients'. A prediction of AD carries ethical dilemmas, but offers a 'future patient' the opportunity to plan their life accordingly. Additionally, a prediction is essential for the development of preventive AD drugs, which—it is to be hoped in the near future—will allow us to reduce the number of AD patients significantly.

Acknowledgements

I would like to thank Dr. Sigrun Roeber for providing the amyloid and tau pictures in Figure 11.1. The work in the author's laboratory is funded by grants from the DFG (SyNergy and FOR2290), BMBF (RiMod-FTD), the Breuer Alzheimer Award, the Helmholtz-Israel program, and the Agency for Innovation by Science and Technology (IWT).

References

Alzheimer, A. 1907. "Über eine eigenartige Erkrankung der Hirnrinde." *Allgemeine Zeitschrift für Psychiatrie und Psychisch-Gerichtliche Medizin* 64:146–8.

Anand, P., and B. Singh. 2013. "A Review on Cholinesterase Inhibitors for Alzheimer's Disease." *Archives of Pharmacal Research* 36(4): 375–99.

Ballard, C., Gauthier, S., Corbett, A., Brayne, C., Aarsland, D., and E. Jones. 2011. "Alzheimer's Disease." *Lancet* 377(9770):1019–31.

Fluhrer, R., Steiner, H., and C. Haass. 2009. "Intramembrane Proteolysis by Signal Peptide Peptidases: A Comparative Discussion of GXGD-Type Aspartyl Proteases." *The Journal of Biological Chemistry* 284(21):13975–9.

Gomez-Isla, T., Price, J. L., McKeel, D. W., Jr., Morris, J. C., Growdon, J. H., and B. T. Hyman. 1996. "Profound Loss of Layer II Entorhinal Cortex Neurons Occurs in Very Mild Alzheimer's Disease." *The Journal of Neuroscience* 16(14):4491–500.

Haass, C., and D. J. Selkoe. 2007. "Soluble Protein Oligomers in Neurodegeneration: Lessons from the Alzheimer's Amyloid Beta-Peptide." *Nature Reviews* 8(2):101–12.

Hardy, J., and D. J. Selkoe. 2002. "The Amyloid Hypothesis of Alzheimer's Disease: Progress and Problems on the Road to Therapeutics." *Science* 297(5580):353–6.

Hebert, L. E., Scherr, P. A., Bienias, J. L., Bennett, D. A., and D. A. Evans. 2003. "Alzheimer Disease in the US Population: Prevalence Estimates Using the 2000 Census." *Archives of Neurology* 60(8):1119–22.

Hendrix, J. A., Finger, B., Weiner, M. W., Frisoni, G. B., Iwatsubo, T., Rowe, C. C., Kim, S. Y., Guinjoan, S. M., Sevlever, G., and M. C. Carrillo. 2015. "The Worldwide Alzheimer's Disease Neuroimaging Initiative: An Update." *Alzheimer's & Dementia: The Journal of the Alzheimer's Association* 11(7):850–9.

Herrup, K. 2010. "Reimagining Alzheimer's Disease—an Age-Based Hypothesis." *The Journal of Neuroscience* 30(50):16755–62.

Holtzman, D. M., Morris, J. C., and A. M. Goate. 2011. "Alzheimer's Disease: The Challenge of the Second Century." *Science Translational Medicine* 3(77):77sr71.

Huang, Y., and L. Mucke. 2012. "Alzheimer Mechanisms and Therapeutic Strategies." *Cell* 148(6):1204–22.

Jonsson, T., Atwal, J. K., Steinberg, S., Snaedal, J., Jonsson, P. V., Bjornsson, S., Stefansson, H., Sulem, P., Gudbjartsson, D., Maloney, J., Hoyte K, Gustafson A, Liu Y, Lu Y, Bhangale T, Graham RR, Huttenlocher J, Bjornsdottir G, Andreassen

OA, Jönsson EG, Palotie A, Behrens TW, Magnusson OT, Kong A, Thorsteinsdottir U, Watts RJ, Stefansson K. 2012. "A Mutation in APP Protects against Alzheimer's Disease and Age-Related Cognitive Decline." *Nature* 488(7409):96–9.

Kanekiyo, T., Xu, H., and G. Bu. 2014. "ApoE and Abeta in Alzheimer's Disease: Accidental Encounters or Partners?" *Neuron* 81(4):740–54.

Karch, C.M., Cruchaga, C., and A.M. Goate. 2014. "Alzheimer's Disease Genetics: From the Bench to the Clinic." *Neuron* 83(1):11–26.

Moulder, K.L., Snider, B.J., Mills, S.L., Buckles, V. D., Santacruz, A. M., Bateman, R. J., and J.C. Morris. 2013. "Dominantly Inherited Alzheimer Network: Facilitating Research and Clinical Trials." *Alzheimer's Research & Therapy* 5(5):48.

Reiman, E.M., Langbaum, J. B., Fleisher, A.S., Caselli, R.J., Chen, K., Ayutyanont, N., Quiroz, Y.T., Kosik, K.S., Lopera, F., and P.N. Tariot. 2011. "Alzheimer's Prevention Initiative: A Plan to Accelerate the Evaluation of Presymptomatic Treatments." *Journal of Alzheimer's Disease* 26(Suppl. 3):321–9.

Schenk, D., Basi, G.S., and M.N. Pangalos. 2012. "Treatment Strategies Targeting Amyloid Beta-Protein." *Cold Spring Harbor Perspectives in Medicine* 2(5):a006387.

Sutphen, C.L., Fagan, A.M., and D.M. Holtzman. 2014. "Progress Update: Fluid and Imaging Biomarkers in Alzheimer's Disease." *Biological Psychiatry* 75(7):520–6.

Vassar, R., Kuhn, P.H., Haass, C., Kennedy, M.E., Rajendran, L., Wong, P.C., and S. F. Lichtenthaler. 2014. "Function, Therapeutic Potential and Cell Biology of BACE Proteases: Current Status and Future Prospects." *Journal of Neurochemistry* 130(1):4–28.

12 Early Detection, Prediction, and Prognosis of Alzheimer's Disease

Simone Lista, Francesco Garaci, Nicola Toschi, and Harald Hampel

Introduction

Recent years have witnessed an increasing understanding of the molecular underpinnings of Alzheimer's disease (AD). The pathogenesis of this complex neurodegenerative disease involves sequentially interacting pathological cascades, including core events (i.e., accumulation of the forty-two-amino-acid-long amyloid β [$A\beta_{1-42}$] peptide into amyloid plaques, and the formation of intraneuronal neurofibrillary tangles), and downstream processes, such as generalized neuroinflammation. These events ultimately cause a deterioration of synaptic integrity, and disrupt normal neural connectivity (Blennow et al. 2006). According to the traditional "amyloid cascade hypothesis," the clearance and degradation of extracellular Aβ is critical for regulating Aβ deposition, and their alterations represent the key event in the pathogenesis of AD (Hardy and Selkoe 2002).

A number of molecular changes have been reported in the AD brain, including, but not limited to, modifications in amyloid precursor protein (APP) metabolism (O'Brien and Wong 2011), tau phosphorylation (Gendron and Petrucelli 2009), mitochondrial dysfunction, increased oxidative stress, neuroinflammatory changes (Verri et al. 2012), lipid alterations (Giannopoulos et al. 2014), membrane lipid dysregulation (Di Paolo and Kim 2011), and abnormal neurotransmission (Francis 2005). Because such alterations are mutually interrelated, a systemic approach is needed to shed more light on the pathogenesis of AD at a complex network level (Noorbakhsh et al. 2009; Kaddurah-Daouk et al. 2013).

Different sets of criteria have been recently proposed for diagnosing AD in subjects with mild cognitive impairment (MCI): the International Working Group (IWG)-1 (Dubois et al. 2010, 2007), IWG-2 (Dubois et al. 2014), and the National Institute of Ageing-Alzheimer Association (NIA-AA) criteria (Albert et al. 2011). All of these criteria encompass the use of biomarkers of AD pathology to increase the reliability of AD as the underlying diagnosis. Nevertheless, they differ both in the definition of MCI and biomarker alterations (Visser et al. 2012). To date, no head-to-head comparison of the different diagnostic criteria has been performed, and the optimal diagnostic approach has yet to be defined. The IWG criteria use the term "prodromal AD" for the diagnosis of AD, and were designed to serve as research criteria. The IWG-1 criteria require episodic memory impairment and at least one abnormal AD biomarker. This biomarker can either be a topographical marker (presence of medial temporal lobe atrophy on magnetic resonance imaging (MRI) or parieto-temporal hypoperfusion on ^{18}F-fluorodeoxyglucose-positron emission tomography [FDG-PET]) or a pathophysiological marker (decreased cerebrospinal fluid [CSF] $A\beta_{1-42}$, increased CSF tau, or increased amyloid PET uptake) (Dubois et al. 2010, 2007). The updated IWG-2 criteria require cognitive impairment in any cognitive domain and (1) decreased CSF $A\beta_{1-42}$ accompanied by increased tau, or (2) increased amyloid PET uptake (Dubois et al. 2014). These criteria define two subcategories: typical prodromal AD (if impairment on a memory test is present) and atypical prodromal AD (if only impairment on a non-memory test is present). Finally, the NIA-AA criteria utilize the term "mild cognitive impairment due to AD," and were intended for both clinical and research uses. They require the presence of cognitive impairment in any cognitive domain, and abnormal amyloid markers (i.e., decreased CSF $A\beta_{1-42}$ or increased amyloid PET uptake), or neuronal injury markers (i.e., presence of medial temporal lobe atrophy on MRI, increased CSF tau, or parieto-temporal hypoperfusion on FDG-PET). According to these criteria, the number of abnormal biomarkers is related to the possibility that MCI is due to AD (Albert et al. 2011).

Biomarker Evidence of the Pathology of Alzheimer's Disease

Recent technical advances have paved the way for a multimodal framework of AD biomarkers, including both biochemical and imaging markers (Hampel et al. 2012; Dubois et al. 2013; Hampel and Lista 2013; Hampel et al. 2014). Current biomarkers of AD are derived from neurogenetics (Zetzsche et al. 2010; Bertram and Hampel 2011; Hampel and Lista 2012), structural/functional/metabolic neuroimaging and neurophysiology (Ewers et al. 2012; Teipel et al. 2013a), neurobiochemistry on biological fluids (Blennow et al. 2012; Rosén et al. 2013), including both CSF (Blennow et al. 2010; Hampel et al. 2010a; Hampel et al. 2010b; Blennow et al. 2015) and blood (plasma/serum) (Gupta et al. 2013; Lista et al. 2013a; Henriksen et al. 2014; O'Bryant et al. 2014; Snyder et al. 2014). However, the diagnostic accuracy of the multimodal approach to AD diagnosis has not been yet established, and should be further evaluated in terms of sensitivity, specificity, and predictive power (Lista and Emanuele 2011; Teipel et al. 2013b; Edwards et al. 2014; Lista et al. 2014). In this scenario, cooperative efforts between industry and regulatory stakeholders, clinicians, researchers, and health care decision-makers are urgently needed (Hampel et al. 2010c; Broich et al. 2011). For a biomarker to mature into a validated and standardized clinical test, it should be feasible, reproducible, and widely available with quality control. Currently, the best established fluid biomarkers for AD include three core CSFs: CSF Aβ_{1-42} (which reflects Aβ plaque formation in the brain), CSF phospho-tau (P-tau; which reflects neurofibrillary tangle pathology in AD), and CSF total-tau (T-tau; a marker of axonal degeneration) markers. All of these biomarkers have been validated against pathology, and have 85–95% sensitivity and specificity for AD in both dementia and prodromal MCI stages (Blennow et al. 2010).

Cerebrospinal Fluid Biomarkers

Diagnostic and Prognostic Performance of CSF Biomarkers

The sensitivity and specificity of a biomarker is crucial for its clinical utility in AD diagnosis. An elevated sensitivity is needed to identify the highest proportion of patients with AD and diminish false negatives. Conversely, a high specificity is essential to maximize the diagnostic accuracy, reduce the number of false positives, and exclude cases unrelated to AD. However, the criteria for a "high" or even "acceptable" accuracy markedly depend on the clinical purposes (e.g., diagnosis, screening, enrichment in clinical trials). The prevalence of a vascular component in AD may also have an impact on the diagnostic performance of currently available biomarkers, requiring investigation in larger studies (Fagan 2014).

AD Dementia

Increased CSF levels of t-tau (Vandermeeren et al. 1993), p-tau (Blennow et al. 1995), and reduced levels of Aβ_{1-42} (Motter et al. 1995) have been well-documented and consistently replicated in AD patients. The sensitivity and specificity of these biomarkers in the dementia phase are 80–95% (Blennow and Hampel 2003). No significant changes in CSF concentrations of these molecules have been reported in other neuropsychiatric disorders, including depression and Parkinson's disease (Blennow et al. 2010). Intriguingly, measurements of CSF p-tau may help distinguish AD from frontotemporal dementia and Lewy-body dementia. Currently available immunoassays for various epitopes of p-tau (e.g., p-tau$_{181}$, p-tau$_{231}$, and p-tau$_{199}$) have similar diagnostic accuracy (Hampel et al. 2004). Notably, all of the CSF biomarkers for AD diagnosis have been validated against pathology (Koopman et al. 2009; Shaw et al. 2009).

Prodromal AD

CSF biomarkers have a high predictive power for detecting prodromal AD in MCI subjects (Blennow and Hampel 2003). Specifically, the combination of the three core CSF biomarkers has a sensitivity of 95% for prodromal AD in MCI (Hansson et al. 2006), and can predict the rate of cognitive decline over time (Snider et al. 2009). The high diagnostic accuracy of CSF biomarkers for prodromal AD has been confirmed in large multicenter studies, including the European Development of Screening guidelines and Criteria for Predementia Alzheimer's disease (DESCRIPA) study (Visser at el. 2009), the Swedish Brain Power (SBP) project (Mattsson et al. 2009), and the US-ADNI (Shaw et al. 2009). Taken together, there is robust evidence to recommend assessment of CSF biomarkers as clinical diagnostic tools for detecting prodromal AD in MCI individuals.

Preclinical AD

Subjects with preclinical AD are defined as those harboring early AD pathology without clinical evidence of cognitive decline. Several studies have investigated whether CSF biomarkers may help identify preclinical AD patients who will progress to clinically overt dementia. Skoog and colleagues (2003) have found that reduced CSF $A\beta_{1-42}$ levels may help predict the development of dementia in initially cognitively healthy 85-year-old individuals (Skoog et al. 2003). These results were confirmed in a population-based study of healthy elderly subjects aged 70–78 years with eight years follow-up (Gustafson et al. 2007), as well as in a clinical cohort of asymptomatic elderly subjects aged 60–94 years (Stomrud et al. 2007). CSF levels of $A\beta_{1-42}$ start declining in the preclinical phase of sporadic AD, before any apparent increase in t-tau or p-tau. With regard to familial AD, Moonis and colleagues (2005) demonstrated that asymptomatic subjects carrying familial AD mutations have both low CSF $A\beta_{1-42}$ and high t-tau levels (Moonis et al. 2005). These findings were subsequently confirmed by Ringman and colleagues (2008), who showed that the full AD pattern of CSF biomarker changes may be detected in mutation carriers years before the onset of symptoms (Ringman et al. 2008). Moreover, Bateman and colleagues (2012) demonstrated that CSF $A\beta_{1-42}$ can start decreasing in familial AD mutation carriers as early as twenty-five years before the clinical onset, whereas increased CSF tau concentrations may be observed at least fifteen years before the appearance of cognitive decline (Bateman et al. 2012). Taken together, these results suggest that CSF biomarkers (especially $A\beta_{1-42}$) are altered several years before the first appearance of cognitive decline. Notably, familial AD mutation carriers display increased CSF $A\beta_{1-42}$ concentrations as of their early 20s (Ringman et al. 2008; Reiman et al. 2012).

Progression from Cognitively Normal Subjects to MCI

Increased CSF ratios of tau/$A\beta_{1-42}$ and p-tau/$A\beta_{1-42}$ have been associated with an increased risk of conversion to MCI in cognitively healthy subjects—notably, 70% of patients with high ratios at a risk of converting to MCI over a three-year period, as compared with a conversion of 10% in those with normal ratios (Fagan et al. 2007). It has been also observed that all converters had an increased tau/Aβ_{1-42} ratio after a mean follow-up of forty-two months (Li et al. 2007). These observations indicate that increased CSF ratios of tau/Aβ_{1-42} and p-tau/Aβ_{1-42} may serve as reliable proxies of amyloid deposition, and denote patients with preclinical AD.

Progression from MCI to AD

Numerous studies have focused on the clinical usefulness of CSF markers in predicting the risk of progression from MCI to AD. In general, reduced Aβ_{1-42} and elevated t-tau and p-tau have similar accuracy in identifying MCI converters; moreover, reduced CSF Aβ_{1-42}/Aβ_{1-40} ratios have been found to predict the risk of progression to AD in subjects with very mild dementia (Brys et al. 2009). A large longitudinal study of MCI subjects (eighteen months follow-up) reported a high tau/Aβ_{1-42} ratio in 90% of MCI subjects who subsequently converted to AD, compared with 10% of those who did not convert (Riemenschneider et al. 2002). Another prospective study reported that the combination of tau with the Aβ_{1-42}/p-tau181 ratio significantly predicted the progression of MCI into AD over a follow-up of four to six years (Hansson et al. 2006). Because the diagnostic accuracy of combining CSF Aβ_{1-42}, t-tau, and p-tau is higher than that of any biomarker alone (Riemenschneider et al. 2002; Hansson et al. 2006; Blennow et al. 2012), a LuminexTM xMAP technology (Luminex Corporation, Austin, TX, USA) multiplex assay has been developed to allow the simultaneous quantification of all of these CSF biomarkers (Olsson et al. 2005). Multicenter studies focusing on this assay yielded promising clinical results (Lewczuk et al. 2008; Mattsson et al. 2009; Shaw et al., 2009).

Blood: Potential Candidate Biomarkers

Although conventional AD biomarkers from CSF are highly accurate, practical barriers to CSF sampling remain a major caveat in routine clinical practice. Because blood is more easily accessible than CSF, research into blood-borne biomarkers of AD has recently gained momentum. The Blood-Based Biomarker Interest Group (BBBIG), an international working group of internationally leading AD blood biomarker scientists from both academia and industry, has recently established relevant collaborative efforts in the field (Henriksen et al. 2014). Although the potential association of plasma Aβ_{1-40} and Aβ_{1-42} concentrations with incipient AD has been repeatedly investigated, definite conclusions are still lacking. Increased Aβ_{1-40} or Aβ_{1-42} levels have been shown to predict the development of AD in some (Mayeux et al. 2003; van Oijen et al. 2006) but not all (Lopez et al. 2008; Sundelof et al. 2008; Hansson et al. 2010) studies. A recent meta-analysis has suggested that a low Aβ_{1-42}/Aβ_{1-40} ratio may predict AD progression, even though no association was observed for single peptides (Koyama et al. 2012).

Although traditional ELISA-based methods for detecting plasma/serum tau lack sufficient analytical sensitivity, novel ultra-sensitive immunoassays have allowed its precise quantification (Randall et al. 2013). Plasma tau levels seem to be increased in AD patients, albeit with a large overlap with cognitively healthy controls (Zetterberg et al. 2013).

Recent advances in mass spectrometry-based technologies has prompted the inspection of the plasma/serum proteome for the discovery of next-generation AD biomarkers (Ray et al. 2011). Most of the novel biomarkers identified to date suggest an involvement of neuroinflammation in AD (Wyss-Coray and Rogers 2012), but validation data are still lacking. This may be due at least in part to methodological issues inherent in plasma/serum proteomics, including pre-analytical variability and lack of standardization in specimens' collection, handling, and processing (Lista et al. 2013b). In this regard, the implementation of rigorous standard operating procedures (SOPs) will be paramount for translation of exploratory proteomic discoveries from bench to bedside (Apweiler et al. 2009). Notably, international consensus guidelines for the preanalytic processing of blood samples for AD biomarker discovery have been recently published (O'Bryant et al. 2014). Finally, global initiatives like the Human Plasma Proteome Project (HPPP), conceptualized by the Human Proteome Organization (HUPO), will be paramount in identifying and validating proteomic-based biomarkers of AD (Omenn et al. 2005).

Neuroimaging Markers

Structural Magnetic Resonance Imaging

Structural MRI allows non-invasive morphometric and volumetric brain evaluation, which can be employed to detect specific regional or focal gray matter (GM) or white matter (WM) alterations. Over the last decade, a large body of literature has used voxel-based morphometry (VBM) (Ashburner 2009) to confirm the mesial temporal areas, the posterior cingulate gyrus, and medial thalamus as the most involved regions in AD patients (Busatto et al. 2008). In addition, longitudinal VBM reports investigating MCI subjects distinguished an MCI-subgroup that progresses to AD from one that remained clinically stable (Whitwell et al. 2007). Several MR studies using VBM have uncovered regional hippocampal and entorhinal cortex atrophy in individuals with mild cognitive impairment (Mungas et al. 2005; Hampel et al. 2008; Lau et al. 2008), and additional VBM studies have identified WM alterations in the angular gyrus, the paracentral region, and postcentral WM when comparing MCI subjects to controls (King et al. 2010; Serra et al. 2010; Teipel et al. 2010). Also, while additional volumetric changes in both AD and MCI subjects may be demonstrated by more advanced MR methods such as Tensor Based (Bossa et al. 2009) and Deformation Based Morphometry (Friese et al. 2010), the majority of such algorithms cannot currently be considered established diagnostic tools (Fox et al. 2011).

WM can be further investigated using Diffusion Tensor Imaging (DTI) (Basser and Jones 2002), which allows probing WM microarchitecture, connectivity, and integrity. DTI has been widely employed in MCI subjects and AD (Bozzali and Cherubini 2007; Chua et al. 2008; Hess et al. 2009). AD-related alterations detected by DTI appeared significantly more diffused than what could be uncovered using VBM, and reached beyond the medial temporal lobe (MTL), demonstrating a higher sensitivity in detecting early ultrastructural WM degeneration (Nucifora et al. 2007; Garaci et al. 2009; Pardini et al. 2009; Friese et al. 2010; Pardini et al. 2012). Also, alterations in DTI-derived indexes have been shown to correlate with disease severity in AD, suggesting that the DTI approach can be used as a non-invasive marker of disease progression (Medina et al. 2006; Fjell et al. 2009; Heo et al. 2009; Ewers et al. 2011; Liu et al. 2011). Although several studies already showed alterations in WM within the parahippocampal gyrus, posterior cingulum, and splenium of the corpus callosum at the MCI stage (Takahashi et al. 2002; Zhang et al. 2007; Chua et al. 2009; Zhuang et al. 2010), most DTI investigations of AD and MCI focus on the entire corpus callosum, uncinate fasciculus, fornix, and the cingulum bundle as the most affected regions (Sexton et al. 2011). Moreover, in agreement with the retrogenesis model (late myelinating fibers would be those more susceptible to neurodegeneration in AD), DTI studies have been able to demonstrate earlier white matter degeneration in specific brain regions such as cortico-cortical prefrontal cortex bundles, inferior longitudinal fasciculus, and temporo-parietal areas (Teipel et al. 2007; Chua et al. 2008; Chua et al. 2009; Stricker et al. 2009).

Functional Magnetic Resonance Imaging

Functional MRI (fMRI) is a non-invasive and safe technique used to map the cortical activity using the blood oxygen level dependent (BOLD) effect. Functional changes in hippocampal and parahippocampal regions which precede clinical AD symptoms have been demonstrated using fMRI (Bookheimer et al. 2000; Gron et al. 2001; Golby et al. 2005; Celone et al. 2006; Hamalainen et al. 2007; Johnson et al. 2006; Petrella et al. 2007; Sperling 2007; Borghesani et al. 2008; Filippini et al. 2009; Sperling et al. 2010), and decreased fMRI activity has been demonstrated in MTL regions when comparing AD patients to controls during episodic encoding tasks (Schwindt and Black 2009). Recently, an increased interest in the potential role of fMRI in the context of early AD diagnosis came from the cortical changes observed in the task-free state ("default mode network") (Celone et al. 2006; Satterthwaite et al. 2007; Pihlajamaki et al. 2008; Fleisher et al. 2009; Pihlajamaki et al. 2010; Pihlajamaki et al. 2011; Vannini et al. 2012). These studies have demonstrated altered default network activity in AD groups (Greicius et al. 2004; Sorg et al. 2007; Wang et al. 2007; Wermke et al.

2008; Greicius et al. 2009), MCI patients, and *APOE* ε4 carriers (Petrella et al. 2011) when compared to controls. Furthermore, a decreased mean level of functional connectivity, as well as diminished fluctuations in the level of synchronization, have been demonstrated in AD (Stam et al. 2005), and the application of a graph theoretical framework to the study of AD and in frontotemporal degeneration has recently revealed the disruption of the small-world-network typical of healthy functional organization (Pievani et al. 2011). To date, only a few studies have demonstrated altered activation after long-term treatment with cholinesterase inhibitors in MCI subjects and AD patients (Rombouts et al. 2002; Saykin et al. 2004; Goekoop et al. 2006; Shanks et al. 2007), and the potential of resting state fMRI studies in elucidating acute and chronic drug-related alterations in function and connectivity has yet to be fully explored (Sperling 2011).

Positron Emission Tomography and Magnetic Resonance Spectroscopy

PET scanning has long been able to reveal reduced blood flow and metabolism in AD, and it can be informative in terms of decision support in the differential diagnosis of some dementia causing diseases (Silverman and Alavi 2005). Early AD has been seen to associate with reduced metabolism of the posterior cingulated cortex (Mosconi et al. 2004) before atrophic changes become detectable through structural MRI (Apostolova et al. 2010). Moreover, Pittsburgh Compound B (PIB)-PET has been shown to closely track the amyloid-beta patterns detected histochemically in post-mortem AD brains (Klunk et al. 2004; Quigley et al. 2011), and Amyloid-PET has been able to detect early frontal lobe degeneration (Mintun et al. 2006; Rowe and Villemagne 2011), hence possibly tracking alterations before any morphometric techniques as well as to associate with early, subclinical cognitive decline (Braskie et al. 2010; Protas et al. 2010). While the high diagnostic sensitivity of PIB-PET is of value in tracking early disease progression, it should be noted that high PIB uptake has been demonstrated in cognitively normal healthy controls (Aizenstein et al. 2008). Although a moderate increase in PIB uptake over time has been seen in AD patients (Grimmer et al. 2010), this finding does not necessarily apply when the assessment extends over several years (Engler et al. 2006; Kadir and Nordberg 2010), suggesting a plateau effect in patients with overt AD. Other recently developed amyloid PET radioligands include florbetapir (^{18}F-AV-45) and ^{18}F-FDDNP, the latter also sensitive to neurofibrillary tangles, whose accumulation has been seen to parallel disease progression, and to amyloid-beta senile plaques in AD patients (Small et al. 2006). In the context of metabolic imaging, the role of Magnetic Resonance Spectroscopy (MRS) in predicting AD risk has not been clearly elucidated (Li and Wahlund 2011); however, reports combining MRS with volumetric approaches (Mueller et al. 2006; Kantarci et al. 2008) suggest a higher specificity and sensitivity in discriminating AD subjects from controls when compared to either approach alone.

Advanced Protocols and Data Processing

It is well known that DTI can only provide limited sensitivity in more complex situations like crossing fiber bundles and/or mixed tissue types (Alexander et al. 2001). This shortcoming is well addressed by novel acquisition and processing strategies like Diffusion Spectrum Imaging (Wedeen et al. 2008), Diffusional Kurtosis Imaging (Hui et al. 2008; Jensen and Helpern 2010; Fieremans et al. 2011), higher order tensor models (Alexander 2005), compartment models (Assaf and Basser 2005; Assaf et al. 2008), and anomalous diffusion (Alexander et al. 2010; De Santis et al. 2011a). These techniques can be optimized in order to fit within the timeframe of a clinical examination (De Santis et al. 2011b), and they have already been successfully employed in characterizing tissue degeneration in several pathologies, including AD (Mintun et al. 2006; Falangola et al. 2008; Iraji et al. 2011; Wang et al. 2011). Further, integrating multimodal imaging protocols into a comprehensive patient characterization strategy is expected to provide better discrimination and staging of AD (Ewers et al. 2011). In this respect, a growing body of literature has appeared which aims to combine multiple modalities (Casanova et al. 2007; Oakes et al. 2007; Yang et al. 2011) in order to formulate more comprehensive hypotheses about structure-function links both in healthy and diseased brains. On the processing side, multimodal data can be integrated and combined with additional patient information (e.g., neuropsychology, genetics, clinical variables) through standard statistical techniques such as independent component analysis (Calhoun et al. 2009), or automatic classifiers, as has been done in AD using support vector machines (Magnin et al. 2009; Haller et al. 2010; Grana et al. 2011). Several studies have begun to deal with AD/healthy controls or MCI/healthy controls classifications (Fan et al. 2008; Hinrichs et al. 2011; Zhang et al. 2011), and a prediction of cognitive decline of MCI subjects has been recently reported (Fan et al. 2008).

Conclusions

Major advances have been achieved in the investigation and development of distinct disease-related biomarkers from different technical modalities for AD during the past two decades. Longitudinal studies have demonstrated that in subjects who ultimately develop cognitive decline and AD dementia, the cognitive signs begin insidiously and, in most cases, evolve gradually. The existence of an AD typical pathological process is supposed to be documented several years or even decades prior to the onset of late stage AD dementia. In light of the findings gained from both neurobiochemical and neuroimaging marker research, AD can be re-conceptualized as a progressive disease evolving from a genetically determined risk state and beginning biochemical alterations in the brain with later appearing cognitive defects, to a condition of cognitive decline with dynamic biomarker concentrations linked to the AD process, to mild, moderate, and severe stages of AD dementia.

Notably, recent progresses in understanding the molecular and cellular mechanisms underlying various paths toward AD pathogenesis have begun to provide new insights for interventions to modify the progression of the disease. The evolving information gained from multidisciplinary basic and translational research has begun to identify new concepts for treatments and distinct classes of therapeutic targets, as well as putative disease-modifying compounds being tested in clinical trials. Biomarkers with a variety of functions will need to be developed and qualified to support optimized clinical trial designs aimed at investigating very early patient populations.

Acknowledgements

S.L. and H.H. are supported by the AXA Research Fund, the Fondation Université Pierre et Marie Curie, and the "*Fondation pour la Recherche sur Alzheimer,*" Paris, France. The research leading to these results has received funding from the program "*Investissements d'avenir*" ANR-10-IAIHU-06.

F.G. and N.T. declare no conflict of interest.

References

Aizenstein, H.J., et al. 2008. "Frequent Amyloid Deposition without Significant Cognitive Impairment among the Elderly." *Archives of Neurology* 65(11):1509–17.

Albert, M.S., et al. 2011. "The Diagnosis of Mild Cognitive Impairment due to Alzheimer's Disease: Recommendations from the National Institute on Aging and Alzheimer's Association Workgroup." *Alzheimer's & Dementia* 7(3):270–9.

Alexander, A.L., et al. 2001. "Analysis of Partial Volume Effects in Diffusion-Tensor MRI." *Magnetic Resonance in Medicine* 45(5):770–80.

Alexander, D. C. 2005. "Multiple-Fiber Reconstruction Algorithms for Diffusion MRI." *Annal of the New York Academy of Sciences* 1064:113–33.

Alexander, D.C., et al. 2010. "Orientationally Invariant Indices of Axon Diameter and Density from Diffusion MRI." *Neuroimage* 52(4):1374–89.

Apostolova, L.G., et al. 2010. "Surface Feature-Guided Mapping of Cerebral Metabolic Changes in Cognitively Normal and Mildly Impaired Elderly." *Molecular Imaging and Biology* 12(2):218–24.

Apweiler, R., et al. 2009. "Approaching Clinical Proteomics: Current State and Future Fields of Application in Fluid Proteomics." *Clinical Chemistry and Laboratory Medicine* 47(6):724–44.

Ashburner, J. 2009. "Computational Anatomy with the SPM Software." *Magnetic Resonance Imaging* 27(8):1163–74.

Assaf, Y., and P.J. Basser. 2005. "Composite Hindered and Restricted Model of Diffusion (CHARMED) MR Imaging of the Human Brain." *Neuroimage* 27(1):48–58.

Assaf, Y., et al. 2008. "AxCaliber: A Method for Measuring Axon Diameter Distribution from Diffusion MRI." *Magnetic Resonance in Medicine* 59(6):1347–54.

Basser, P.J., and D.K. Jones. 2002. "Diffusion-Tensor MRI: Theory, Experimental Design and Data Analysis—a Technical Review." *NMR in Biomedicine* 15(7–8):456–67.

Bateman, R.J., et al. 2012. "Clinical and Biomarker Changes in Dominantly Inherited Alzheimer's Disease." *The New England Journal of Medicine* 367(9):795–804.

Bertram, L., and H. Hampel. 2011. "The Role of Genetics for Biomarker Development in Neurodegeneration." *Progress in Neurobiology* 95(4):501–14.

Blennow, K., et al. 1995. "Tau Protein in Cerebrospinal Fluid: A Biochemical Marker for Axonal Degeneration in Alzheimer Disease." *Molecular and Chemical Neuropathology* 26(3):231–45.

Blennow, K., et al. 2006. "Alzheimer's Disease." *Lancet* 368(9533):387–403.

Blennow, K., et al. 2010. "Cerebrospinal Fluid and Plasma Biomarkers in Alzheimer Disease." *Nature Reviews Neurology* 6(3):131–44.

Blennow, K., et al. 2012. "Fluid Biomarkers in Alzheimer Disease." *Cold Spring Harbor Perspectives in Medicine* 2(9):a006221.

Blennow, K., et al. 2015. "Clinical Utility of Cerebrospinal Fluid Biomarkers in the Diagnosis of Early Alzheimer's Disease." *Alzheimer's & Dementia* 11(1):58–69.

Blennow, K., and H. Hampel. 2003. "Cerebrospinal Fluid Markers for Incipient Alzheimer's Disease." *Lancet Neurology* 2(10):605–13.

Bookheimer, S. Y., et al. 2000. "Patterns of Brain Activation in People at Risk for Alzheimer's Disease." *The New England Journal of Medicine* 343(7):450–6.

Borghesani, P. R., et al. 2008. "Altered Medial Temporal Lobe Responses During Visuospatial Encoding in Healthy APOE*4 Carriers." *Neurobiology of Aging* 29(7):981–91.

Bossa, M. N., et al. 2009. "Tensor-based Morphometry with Mappings Parameterized by Stationary Velocity Fields in Alzheimer's Disease Neuroimaging Initiative." *Medical Image Computing Computer Assisted Intervention* 12(Pt 2):240–7.

Bozzali, M., and A. Cherubini. 2007. "Diffusion Tensor MRI to Investigate Dementias: A Brief Review." *Magnetic Resonance Imaging* 25(6):969–77.

Braskie, M. N., et al. 2010. "Plaque and Tangle Imaging and Cognition in Normal Aging and Alzheimer's Disease." *Neurobiology of Aging* 31(10):1669–78.

Broich, K., et al. 2011. "Biomarkers in Clinical Trials for Neurodegenerative Diseases: Regulatory Perspectives and Requirements." *Progress in Neurobiology* 95(4):498–500.

Brys, M., et al. 2009. "Prediction and Longitudinal Study of CSF Biomarkers in Mild Cognitive Impairment." *Neurobiology of Aging* 30(5):682–90.

Busatto, G. F., et al. 2008. "Voxel-based Morphometry in Alzheimer's Disease." *Expert Review of Neurotherapeutics* 8(11):1691–702.

Calhoun, V. D., et al. 2009. "A Review of Group ICA for fMRI Data and ICA for Joint Inference of Imaging, Genetic, and ERP Data." *Neuroimage* 45(Suppl. 1):S163–72.

Casanova, R., et al. 2007. "Biological Parametric Mapping: A Statistical Toolbox for Multimodality Brain Image Analysis." *Neuroimage* 34(1):137–43.

Celone, K. A., et al. 2006. "Alterations in Memory Networks in Mild Cognitive Impairment and Alzheimer's Disease: An Independent Component Analysis." *Journal of Neuroscience* 26(40):10222–31.

Chua, T. C., et al. 2008. "Diffusion Tensor Imaging in Mild Cognitive Impairment and Alzheimer's Disease: A Review." *Current Opinion in Neurology* 21(1):83–92.

Chua, T. C., et al. 2009. "Diffusion Tensor Imaging of the Posterior Cingulate Is a Useful Biomarker of Mild Cognitive Impairment." *The American Journal of Geriatric Psychiatry* 17(7):602–13.

De Santis, S., et al. 2011a. "Anisotropic Anomalous Diffusion Assessed in the Human Brain by Scalar Invariant Indices." *Magnetic Resonance in Medicine* 65(4):1043–52.

De Santis, S., et al. 2011b. "Non-Gaussian Diffusion Imaging: A Brief Practical Review." *Magnetic Resonance Imaging* 29(10):1410–16.

Di Paolo, G., and T.W. Kim. 2011. "Linking Lipids to Alzheimer's Disease: Cholesterol and Beyond." *Nature Reviews Neuroscience* 12(5):284–96.

Dubois, B., et al. 2007. "Research Criteria for the Diagnosis of Alzheimer's Disease: Revising the NINCDS-ADRDA Criteria." *Lancet Neurology* 6(8):734–46.

Dubois, B., et al. 2010. "Revising the Definition of Alzheimer's Disease: A New Lexicon." *Lancet Neurology* 9(11):1118–27.

Dubois, B., et al. 2013. "Alzheimer Disease: From Biomarkers to Diagnosis." *Revue Neurologique (Paris)* 169(10):744–51.

Dubois, B., et al. 2014. "Advancing Research Diagnostic Criteria for Alzheimer's Disease: The IWG-2 Criteria." *Lancet Neurology* 13(6):614–29.

Edwards, M., et al. 2014. "Combining Select Neuropsychological Assessment with Blood-Based Biomarkers to Detect Mild Alzheimer's Disease: A Molecular Neuropsychology Approach." *Journal of Alzheimer's Disease* 42(2):635–40.

Engler, H., et al. 2006. "Two-Year Follow-Up of Amyloid Deposition in Patients with Alzheimer's Disease." *Brain* 129(Pt 11):2856–66.

Ewers, M., et al. 2011. "Staging Alzheimer's Disease Progression with multimodality Neuroimaging." *Progress in Neurobiology* 95(4):535–46.

Ewers, M., et al. 2012. "Neuroimaging Markers for the Prediction and Early Diagnosis of Alzheimer's Disease Dementia." *Trends in Neuroscience* 34(8):430–42.

Fagan, A.M. 2014. "CSF Biomarkers of Alzheimer's Disease: Impact on Disease Concept, Diagnosis, and Clinical Trial Design." *Advances in Geriatrics*.

Fagan, A. M., et al. 2007. "Cerebrospinal Fluid Tau/Beta-Amyloid(42) Ratio as a Prediction of Cognitive Decline in Nondemented Older Adults." *Archives of Neurology* 64(3):343–9.

Falangola, M.F., et al. 2008. "Age-Related Non-Gaussian Diffusion Patterns in the Prefrontal Brain." *Journal of Magnetic Resonance Imaging* 28(6):1345–50.

Fan, Y., et al. 2008. "Spatial Patterns of Brain Atrophy in MCI Patients, Identified via High-Dimensional Pattern Classification, Predict Subsequent Cognitive Decline." *Neuroimage* 39(4):1731–43.

Fieremans, E., et al. 2011. "White Matter Characterization with Diffusional Kurtosis Imaging." *Neuroimage* 58(1):177–88.

Filippini, N., et al. 2009. "Distinct Patterns of Brain Activity in Young Carriers of the APOE-epsilon4 Allele." *Proceedings of the National Academy of Sciences USA* 106(17):7209–14.

Fjell, A.M., et al. 2009. "Mini-Mental State Examination Is Sensitive to Brain Atrophy in Alzheimer's Disease." *Dementia and Geriatric Cognitive Disorders* 28(3):252–8.

Fleisher, A.S., et al. 2009. "Resting-State Bold Networks versus Task-Associated Functional MRI for Distinguishing Alzheimer's Disease Risk Groups." *Neuroimage* 47(4):1678–90.

Fox, N.C., et al. 2011. "Algorithms, Atrophy and Alzheimer's Disease: Cautionary Tales for Clinical Trials." *Neuroimage* 57(1):15–18.

Francis, P.T. 2005. "The Interplay of Neurotransmitters in Alzheimer's Disease." *CNS Spectr* 10(11 Suppl 18):6–9.

Friese, U., et al. 2010. "Diagnostic Utility of Novel MRI-Based Biomarkers for Alzheimer's Disease: Diffusion Tensor Imaging and Deformation-Based Morphometry." *Journal of Alzheimer's Diseases* 20(2):477–90.

Garaci, F.G., et al. 2009. "Optic Nerve and Optic Radiation Neurodegeneration in Patients with Glaucoma: In Vivo Analysis with 3-t Diffusion-Tensor MR Imaging." *Radiology* 252(2):496–501.
Gendron, T.F., and L. Petrucelli. 2009. "The Role of Tau in Neurodegeneration." *Molecular Neurodegeneration* 4:13.
Giannopoulos, P.F., et al. 2014. "Novel Lipid Signaling Pathways in Alzheimer's Disease Pathogenesis." *Biochemical Pharmacology* 88(4):560–64.
Goekoop, R., et al. 2006. "Cholinergic Challenge in Alzheimer Patients and Mild Cognitive Impairment Differentially Affects Hippocampal Activation—a Pharmacological fMRI Study." *Brain* 129(Pt 1):141–57.
Golby, A., et al. 2005. "Memory Encoding in Alzheimer's Disease: An fMRI Study of Explicit and Implicit Memory." *Brain* 128(Pt 4):773–87.
Grana, M., et al. 2011. "Computer Aided Diagnosis System for Alzheimer Disease Using Brain Diffusion Tensor Imaging Features Selected by Pearson's Correlation." *Neuroscience Letters* 502(3):225–9.
Greicius, M.D., et al. 2004. "Default-Mode Network Activity Distinguishes Alzheimer's Disease from Healthy Aging: Evidence from Functional fMRI." *Proceedings of National Academy Sciences USA* 101(13):4637–42.
Greicius, M.D., et al. 2009. "Resting-State Functional Connectivity Reflects Structural Connectivity in the Default Mode Network." *Cerebral Cortex* 19(1):72–8.
Grimmer, T., et al. 2010. "Progression of Cerebral Amyloid Load Is Associated with the Apolipoprotein E ε4 Genotype in Alzheimer's Disease." *Biological Psychiatry* 68(10):879–84.
Gron, G., et al. 2001. "Hippocampal Activations during Repetitive Learning and Recall of Geometric Patterns." *Learning & Memory* 8(6):336–45.
Gupta, V.B., et al. 2013. "Multiplex Biomarkers in Blood." *Alzheimer's Research & Therapy* 5(3):31.
Gustafson, D.R., et al. 2007. "Cerebrospinal Fluid Beta-Amyloid 1–42 Concentration May Predict Cognitive Decline in Older Women." *Journal of Neurology, Neurosurgery & Psychiatry* 78(5):461–4.
Haller, S., et al. 2010. "Individual Prediction of Cognitive Decline in Mild Cognitive Impairment using Support Vector Machine-Based Analysis of Diffusion Tensor Imaging Data." *Journal of Alzheimer's Disease* 22(1):315–27.
Hamalainen, A., et al. 2007. "Increased fMRI Responses During Encoding in Mild Cognitive Impairment." *Neurobiology of Aging* 28(12):1889–903.
Hampel, H., et al. 2004. "Value of CSF Beta-Amyloid1–42 and Tau as Predictors of Alzheimer's Disease in Patients with Mild Cognitive Impairment." *Molecular Psychiatry* 9(7):705–10.
Hampel, H., et al. 2008. "Core Candidate Neurochemical and Imaging Biomarkers of Alzheimer's Disease." *Alzheimer's & Dementia* 4(1):38–48.
Hampel, H., et al. 2010a. "Biological Markers of Amyloid Beta-Related Mechanisms in Alzheimer's Disease." *Experimental Neurology* 223(2):334–46.
Hampel, H., et al. 2010b. "Total and Phosphorylated Tau Protein as Biological Markers of Alzheimer's Disease." *Experimental Gerontology* 45(1):30–40.
Hampel, H., et al. 2010c. "Biomarkers for Alzheimer's Disease: Academic, Industry and Regulatory Perspectives." *Nature Reviews Drug Discovery* 9(7):560–74.
Hampel, H., et al. 2012. "Development of Biomarkers to Chart all Alzheimer's Disease Stages: The Royal Road to Cutting the Therapeutic Gordian Knot." *Alzheimer's & Dementia* 8(4):312–36.

Hampel, H., et al. 2014. "Perspective on Future Role of Biological Markers in Clinical Therapy Trials of Alzheimer's Disease: A Long-Range Point of View Beyond 2020." *Biochemical Pharmacology* 88(4):426–49.

Hampel, H., and S. Lista. 2012. "Alzheimer Disease: From Inherited to Sporadic AD-Crossing the Biomarker Bridge." *Nature Reviews Neurology* 8(11):598–600.

Hampel, H., and S. Lista. 2013. "Use of Biomarkers and Imaging to Assess Pathophysiology, Mechanisms of Action and Target Engagement." *The Journal of Nutrition Health and Aging* 17(1):54–63.

Hansson, O., et al. 2006. "Association between CSF Biomarkers and Incipient Alzheimer's Disease in Patients with Mild Cognitive Impairment: A Follow-Up Study." *Lancet Neurology* 5(3):228–34.

Hansson, O., et al. 2010. "Evaluation of Plasma Abeta(40) and Abeta(42) as Predictors of Conversion to Alzheimer's Disease in Patients with Mild Cognitive Impairment." *Neurobiology of Aging* 31(3):357–67.

Hardy, J., and D.J. Selkoe. 2009. "The Amyloid Hypothesis of Alzheimer's Disease: Progress and Problems on the Road to Therapeutics." *Science* 297(5580):353–6.

Henriksen, K., et al. 2014. "The Future of Blood-Based Biomarkers for Alzheimer's Disease." *Alzheimer's & Dementia* 10(1):115–31.

Heo, J. H., et al. 2009. "White Matter Hyperintensities and Cognitive Dysfunction in Alzheimer Disease." *Journal of Geriatric Psychiatry and Neurology* 22(3):207–12.

Hess, C.P. 2009. "Update on Diffusion Tensor Imaging in Alzheimer's Disease." *Magnetic Resonance Imaging Clinics of North America* 17(2):215–24.

Hinrichs, C., et al. 2011. "Predictive Markers for AD in a Multi-Modality Framework: An Analysis of MCI Progression in the ADNI Population." *Neuroimage* 55(2):574–89.

Hui, E.S., et al. 2008. "Advanced MR Diffusion Characterization of Neural Tissue using Directional Diffusion Kurtosis Analysis." *Conference Proceedings: . . . Annual International Conference of the IEEE Engineering in Medicine and Biology Society* 2008:3941–4.

Human Proteome Organization. 2005. "Plasma Proteome Project." Available at https://www.hupo.org/plasma-proteome-project/ [accessed March 14, 2016].

Iraji, A., et al. 2011. "Diffusion Kurtosis Imaging Discriminates Patients with White Matter Lesions from Healthy Subjects." *Conference Proceedings: . . . Annual International Conference of the IEEE Engineering in Medicine Biology and Society* 2011:2796–9.

Jensen, J.H., and J. A. Helpern. 2010. "Progress in Diffusion-Weighted Imaging: Concepts, Techniques and Applications to the Central Nervous System." *NMR in Biomedicine* 23(7):659–60.

Johnson, S.C., et al. 2006. "Activation of Brain Regions Vulnerable to Alzheimer's Disease: The Effect of Mild Cognitive Impairment." *Neurobiology of Aging* 27(11):1604–12.

Kaddurah-Daouk, R., et al. 2013. "Alterations in Metabolic Pathways and Networks in Alzheimer's Disease." *Translational Psychiatry* 3:e244.

Kadir, A., and A. Nordberg. 2010. "Target-Specific PET Probes for Neurodegenerative Disorders Related to Dementia." *Journal of Nuclear Medicine* 51(9):1418–30.

Kantarci, K., et al. 2008. "Hippocampal Volumes, Proton Magnetic Resonance Spectroscopy Metabolites, and Cerebrovascular Disease in Mild Cognitive Impairment Subtypes." *Archives of Neurology* 65(12):1621–8.

King, R. D., et al. 2010. "Fractal Dimension Analysis of the Cortical Ribbon in Mild Alzheimer's Disease." *Neuroimage* 53(2):47147–9.

Klunk, W. E., et al. 2004. "Imaging Brain Amyloid in Alzheimer's Disease with Pittsburgh Compound-B." *Annals of Neurology* 55(3):306–19.

Koopman, K., et al. 2009. "Improved Discrimination of Autopsy-Confirmed Alzheimer's Disease (AD) from Non-AD Dementias Using CSF P-tau(181P)." *Neurochemistry International* 55(4):214–18.

Koyama, A., et al. 2012. "Plasma Amyloid-Beta as a Predictor of Dementia and Cognitive Decline: A Systematic Review and Meta-Analysis." *Archives of Neurology* 69(7):824–31.

Lau, J. C., et al. 2008. "Longitudinal Neuroanatomical Changes Determined by Deformation-based Morphometry in a Mouse Model of Alzheimer's Disease." *Neuroimage* 42(1):19–27.

Lewczuk, P., et al. 2008. "Multiplexed Quantification of Dementia Biomarkers in the CSF of Patients with Early Dementias and MCI: A Multicenter Study." *Neurobiology of Aging* 29(6):812–18.

Li, G., et al. 2007. "CSF tau/Abeta42 Ratio for Increased Risk of Mild Cognitive Impairment: A Follow-Up Study." *Neurology* 69(7):631–9.

Li, T. Q., and L. O. Wahlund. 2011. "The Search for Neuroimaging Biomarkers of Alzheimer's Disease with Advanced MRI Techniques." *Acta Radiologica* 52(2):211–22.

Lista, S., and E. Emanuele. 2011. "Role of Amyloid β1–42 and Neuroimaging Biomarkers in Alzheimer's Disease." *Biomarkers in Medicine* 5(4):411–13.

Lista, S., et al. 2013a. "Blood and Plasma-based Proteomic Biomarker Research in Alzheimer's Disease." *Progress in Neurobiology* 101–2:1–17.

Lista, S., et al. 2013b. "Biological and Methodical Challenges of Blood-Based Proteomics in the Field of Neurological Research." *Progress in Neurobiology* 101–2:18–34.

Lista, S., et al. 2014. "CSF Aβ1–42 Combined with Neuroimaging Biomarkers in the Early Detection, Diagnosis and Prediction of Alzheimer's Disease." *Alzheimer's & Dementia* 10(3):381–92.

Liu, Y., et al. 2011. "Diffusion Tensor Imaging and Tract-Based Spatial Statistics in Alzheimer's Disease and Mild Cognitive Impairment." *Neurobiology of Aging* 32(9):1558–71.

Lopez, O. L., et al. 2008. "Plasma Amyloid Levels and the Risk of AD in Normal Subjects in the Cardiovascular Health Study." *Neurology* 70(19):1664–71.

Magnin, B., et al. 2009. "Support Vector Machine-based Classification of Alzheimer's Disease from Whole-Brain Anatomical MRI." *Neuroradiology* 51(2):73–83.

Mattsson, N., et al. 2009. "CSF Biomarkers and Incipient Alzheimer Disease in Patients with Mild Cognitive Impairment." *JAMA* 302(4):385–93.

Mayeux, R., et al. 2003. "Plasma A[beta]40 and A[beta]42 and Alzheimer's Disease: Relation to Age, Mortality, and Risk." *Neurology* 61(9):1185–90.

Medina, D., et al. 2006. "White Matter Changes in Mild Cognitive Impairment and AD: A Diffusion Tensor Imaging Study." *Neurobiology of Aging* 27(5):663–72.

Mintun, M. A., et al. 2006. "[11C]PIB in a Nondemented Population: Potential Antecedent Marker of Alzheimer Disease." *Neurology* 67(3):446–52.

Moonis, M., et al. 2005. "Familial Alzheimer Disease: Decreases in CSF Abeta42 Levels Precede Cognitive Decline." *Neurology* 65(2):323–5.

Mosconi, L., et al. 2004. "Magnetic Resonance and PET Studies in the Early Diagnosis of Alzheimer's Disease." *Expert Review of Neurotherapeutics* 4(5):831–49.

Motter, R., et al. 1995. "Reduction of Beta-Amyloid Peptide42 in the Cerebrospinal Fluid of Patients with Alzheimer's Disease." *Annals of Neurology* 38(4):643–8.

Mueller, S.G., et al. 2006. "Evaluation of Treatment Effects in Alzheimer's and Other Neurodegenerative Diseases by MRI and MRS." *NMR in Biomedicine* 19(6):655–68.

Mungas, D., et al. 2005. "Longitudinal Volumetric MRI Change and Rate of Cognitive Decline." *Neurology* 65(4):565–71.

Noorbakhsh, F., et al. 2009. "Deciphering Complex Mechanisms in Neurodegenerative Diseases: The Advent of Systems Biology." *Trends in Neuroscience* 32(2):88–100.

Nucifora, P.G., et al. 2007. "Diffusion-Tensor MR Imaging and Tractography: Exploring Brain Microstructure and Connectivity." *Radiology* 245(2):367–84.

Oakes, T. R., et al. 2007. "Integrating VBM into the General Linear Model with Voxelwise Anatomical Covariates." *Neuroimage* 34(2):500–8.

O'Brien, R.J., and P. C. Wong. 2011. "Amyloid Precursor Protein Processing and Alzheimer's Disease." *Annual Review of Neuroscience* 34:185–204.

O'Bryant, S.E., et al. 2014. "Guidelines for the Standardization of Preanalytic Variables for Blood-based Biomarker Studies in Alzheimer's Disease Research." *Alzheimer's & Dementia* 11(5):549–60.Olsson, A., et al. 2005. "Simultaneous Measurement of B-Amyloid(1–42), Tau and Phosphory-Lated Tau (Thr181) in Cerebrospinal Fluid by the xMAP Technology." *Clinical Chemistry* 51(2):336–45.

Omenn, G. S., et al. 2005. "Overview of the HUPO Plasma Proteome Project: Results from the Pilot Phase with 35 Collaborating Laboratories and Multiple Analytical Groups, Generating a Core Dataset of 3020 Proteins and a Publicly-Available Database." *Proteomics* 5(13):3226–45.

Pardini, M., et al. 2009. "White Matter Reduced Streamline Coherence in Young Men with Autism and Mental Retardation." *European Journal of Neurology* 16(11):1185–90.

Pardini, M., et al. 2012. "Long-Term Cognitive and Behavioral Therapies, Combined with Augmentative Communication, Are Related to Uncinate Fasciculus Integrity in Autism." *Journal of Autism and Developmental Disorders* 42(4):585–92.

Petrella, J.R., et al. 2007. "Cortical Deactivation in Mild Cognitive Impairment: High-Field-Strength Functional MR Imaging." *Radiology* 245(1):224–35.

Petrella, J.R., et al. 2011. "Default Mode Network Connectivity in Stable vs. Progressive Mild Cognitive Impairment." *Neurology* 76(6):511–17.

Pievani, M., et al. 2011. "Functional Network Disruption in the Degenerative Dementias." *Lancet Neurology* 10(9):829–43.

Pihlajamaki, M., et al. 2008. "Impaired Medial Temporal Repetition Suppression Is related to Failure of Parietal Deactivation in Alzheimer Disease." *American Journal of Geriatric Psychiatry* 16(4):283–92.

Pihlajamaki, M., et al. 2010. "Evidence of Altered Posteromedial Cortical FMRI Activity in Subjects at Risk for Alzheimer Disease." *Alzheimer Disease & Associated Disorders* 24(1):28–36.

Pihlajamaki, M., et al. 2011. "Failure of Repetition Suppression and Memory Encoding in Aging and Alzheimer's Disease." *Brain Imaging and Behavior* 5(1):36–44.

Protas, H.D., et al. 2010. "FDDNP Binding using MR Derived Cortical Surface Maps." *Neuroimage* 49(1):240–48.

Quigley, H., et al. 2011. "PET Imaging of Brain Amyloid in Dementia: A Review." *International Journal of Geriatric Psychiatry* 26(10):991–9.

Randall, J., et al. 2013. "Tau Proteins in Serum Predict Neurological Outcome after Hypoxic Brain Injury from Cardiac Arrest: Results of a Pilot Study." *Resuscitation* 84(3):351–6.

Ray, S., et al. 2011. "Proteomic Technologies for the Identification of Disease Biomarkers in Serum: Advances and Challenges Ahead." *Proteomics* 11(11):2139–61.

Reiman, E.M., et al. 2012. "Brain Imaging and Fluid Biomarker Analysis in Young Adults at Genetic Risk for Autosomal Dominant Alzheimer's Disease in the Presenilin 1 E280A Kindred: A Case—Control Study." *Lancet Neurology* 11(12):1048–56.

Riemenschneider, M., et al. 2002. "Cerebrospinal Fluid Tau and Beta-Amyloid 42 Proteins Identify Alzheimer Disease in Subjects with Mild Cognitive Impairment." *Archives of Neurology* 59(11):1729–34.

Ringman, J.M., et al. 2008. "Biochemical Markers in Persons with Preclinical Familial Alzheimer Disease." *Neurology* 71(2):85–92.

Rombouts, S.A., et al. 2002. "Alterations in Brain Activation during Cholinergic Enhancement with Rivastigmine in Alzheimer's Disease." *Journal of Neurology, Neurosurgery & Psychiatry* 73(6):665–71.

Rosén, C., et al. 2013. "Fluid Biomarkers in Alzheimer's Disease—Current Concepts." *Molecular Neurodegeneration* 8:20.

Rowe, C.C., and V. L. Villemagne. 2011. "Brain Amyloid Imaging." *Journal of Nuclear Medicine* 52(11):1733–40.

Satterthwaite, T.D., et al. 2007. "Dissociable But Inter-Related Systems of Cognitive Control and Reward during Decision Making: Evidence from Pupillometry and Event-Related fMRI." *Neuroimage* 37(3):1017–31.

Saykin, A.J., et al. 2004. "Cholinergic Enhancement of Frontal Lobe Activity in Mild Cognitive Impairment." *Brain* 127(Pt 7):1574–83.

Schwindt, G.C., and S.E. Black. 2009. "Functional Imaging Studies of Episodic Memory in Alzheimer's Disease: A Quantitative Meta-Analysis." *Neuroimage* 45(1):181–90.

Serra, L., et al. 2010. "Grey and White Matter Changes at Different Stages of Alzheimer's Disease." *Journal of Alzheimer's Disease* 19(1):147–59.

Sexton, C.E., et al. 2011. "A Meta-Analysis of Diffusion Tensor Imaging in Mild Cognitive Impairment and Alzheimer's Disease." *Neurobiology of Aging* 32(12):2322 e5–18.

Shanks, M.F., et al. 2007. "Regional Brain Activity after Prolonged Cholinergic Enhancement in Early Alzheimer's Disease." *Magnetic Resonance Imaging* 25(6): 848–59.

Shaw, L.M., et al. 2009. "Cerebrospinal Fluid Biomarker Signature in Alzheimer's Disease Neuroimaging Initiative Subjects." *Annals of Neurology* 65(4):403–13.

Silverman, D. H., and A. Alavi. 2005. "PET Imaging in the Assessment of Normal and Impaired Cognitive Function." *Radiologic Clinics of North America* 43(1):67–77.

Skoog, I., et al. 2003. "Cerebrospinal Fluid Beta-Amyloid 42 Is Reduced Before the Onset of Sporadic Dementia: A Population-Based Study in 85-Year-Olds." *Dementia and Geriatric Cognitive Disorders* 15(3):169–76.

Small, G.W., et al. 2006. "PET of Brain Amyloid and Tau in Mild Cognitive Impairment." *The New England Journal of Medicine* 355(25):2652–63.

Snider, B.J., et al. 2009. "Cerebrospinal Fluid Biomarkers and Rate of Cognitive Decline in Very Mild Dementia of the Alzheimer Type." *Archives of Neurology* 66(5):638–45.

Snyder, H.M., et al. 2014. "Developing Novel Blood-Based Biomarkers for Alzheimer's Disease." *Alzheimer's & Dementia* 10(1):109–14.

Sorg, C., et al. 2007. "Selective Changes of Resting-State Networks in Individuals at Risk for Alzheimer's Disease." *Proceeding of the National Academy of Sciences USA* 104(47):18760–5.

Sperling, R. 2007. "Functional MRI Studies of Associative Encoding in Normal Aging, Mild Cognitive Impairment, and Alzheimer's Disease." *Annals of the New York Academy of Sciences* 1097:146–55.

Sperling, R. 2011. "Potential of Functional MRI as a Biomarker in Early Alzheimer's Disease." *Neurobiology of Aging* 32(Suppl 1):37–43.

Sperling, R. A., et al. 2010. "Functional Alterations in Memory Networks in Early Alzheimer's Disease." *Neuromolecular Medicine* 12(1):27–43.

Stam, C.J., et al. 2005. "Disturbed Fluctuations of Resting State EEG Synchronization in Alzheimer's Disease." *Clinical Neurophysiology* 116(3):708–15.

Stomrud, E., et al. 2007. "Cerebrospinal Fluid Biomarkers Predict Decline in Subjective Cognitive Function over 3 Years in Healthy Elderly." *Dementia and Geriatric Cognitive Disorders* 24(2):118–24.

Stricker, N.H., et al. 2009. "Decreased White Matter Integrity in Late-Myelinating Fiber Pathways in Alzheimer's Disease Supports Retrogenesis." *Neuroimage* 45(1):10–16.

Sundelof, J., et al. 2008. "Plasma Beta Amyloid and the Risk of Alzheimer Disease and Dementia in Elderly Men: A Prospective, Population-Based Cohort Study." *Archives of Neurology* 65(2):256–63.

Takahashi, S., et al. 2002. "Selective Reduction of Diffusion Anisotropy in White Matter of Alzheimer Disease Brains Measured by 3.0 Tesla Magnetic Resonance Imaging." *Neuroscience Letters* 332(1):45–8.

Teipel, S.J., et al. 2007. "Multivariate Network Analysis of Fiber Tract Integrity in Alzheimer's Disease." *Neuroimage* 34(3):985–95.

Teipel, S. J., et al. 2010. "Multicentre Variability of MRI-based Medial Temporal Lobe Volumetry in Alzheimer's Disease." *Psychiatry Research* 182(3):244–50.

Teipel, S. J., et al. 2013a. "Relevance of Magnetic Resonance Imaging for Early Detection and Diagnosis of Alzheimer Disease." *Medical Clinics of North America* 97(3):399–424.

Teipel, S. J., et al. 2013b. "Perspectives for Multimodal Neurochemical and Imaging Biomarkers in Alzheimer's Disease." *Journal of Alzheimer's Disease* 33(Suppl. 1):S329–47.

Vandermeeren, M., et al. 1993. "Detection of Tau Proteins in Normal and Alzheimer's Disease Cerebrospinal Fluid with a Sensitive Sandwich Enzyme-Linked Immunosorbent Assay." *Journal of Neurochemistry* 61(5):1828–34.

Vannini, P., et al. 2012. "Age and Amyloid-Related Alterations in Default Network Habituation to Stimulus Repetition." *Neurobiology of Aging* 33(7):1237–52.

van Oijen, M., et al. 2006. "Plasma Abeta(1–40) and Abeta(1–42) and the Risk of Dementia: A Prospective Case—Cohort Study." *Lancet Neurology* 5(8):655–60.

Verri, M., et al. 2012. "Mitochondrial Alterations, Oxidative Stress and Neuroinflammation in Alzheimer's Disease." *International Journal of Immunopathology and Pharmacology* 25(2):345–53.

Visser, P. J., et al. 2009. "Prevalence and Prognostic Value of CSF Markers of Alzheimer's Disease Pathology in Patients with Subjective Cognitive Impairment or Mild Cognitive Impairment in the DESCRIPA study: A Prospective Cohort Study." *Lancet Neurology* 8(7):619–27.

Visser, P. J., et al. 2012. "Comparison of International Working Group Criteria and National Institute on Aging-Alzheimer's Association Criteria for Alzheimer's Disease." *Alzheimer's & Dementia* 8(6):560–3.

Wang, J. J., et al. 2011. "Parkinson Disease: Diagnostic Utility of Diffusion Kurtosis Imaging." *Radiology* 261(1):210–17.

Wang, K., et al. 2007. "Altered Functional Connectivity in Early Alzheimer's Disease: A Resting-State fMRI study." *Human Brain Mapping* 28(10):967–78.

Wedeen, V. J., et al. 2008. "Diffusion Spectrum Magnetic Resonance Imaging (DSI) Tractography of Crossing Fibers." *Neuroimage* 41(4):1267–77.

Wermke, M., et al. 2008. "A New Integrative Model of Cerebral Activation, Deactivation and Default Mode Function in Alzheimer's Disease." *European Journal of Nuclear Medicine Molecular Imaging* 35(Suppl 1):S12–24.

Whitwell, J. L., et al. 2007. "3D Maps From Multiple MRI Illustrate Changing Atrophy Patterns as Subjects Progress from Mild Cognitive Impairment to Alzheimer's Disease." *Brain* 130(Pt 7):1777–86.

Wyss-Coray, T., and J. Rogers. 2012. "Inflammation in Alzheimer Disease—a Brief Review of the Basic Science and Clinical Literature." *Cold Spring Harbor Perspectives in Medicine* 2(1). doi: 10.1101/cshperspect.a006346.

Yang, X., et al. 2011. "Biological Parametric Mapping with Robust and Non-Parametric Statistics." *Neuroimage* 57(2):423–30.

Zetterberg, H., et al. 2013. "Plasma Tau Levels in Alzheimer's Disease." *Alzheimer's Research & Therapy* 5(2):9.

Zetzsche, T., et al. 2010. "Advances and Perspectives from Genetic Research: Development of Biological Markers in Alzheimer's Disease." *Expert Review of Molecular Diagnostics* 10(5):667–90.

Zhang, D., et al. 2011. "Multimodal Classification of Alzheimer's Disease and Mild Cognitive Impairment." *Neuroimage* 55(3):856–67.

Zhang, Y., et al. 2007. "Diffusion Tensor Imaging of Cingulum Fibers in Mild Cognitive Impairment and Alzheimer Disease." *Neurology* 68(1):13–19.

Zhuang, L., et al. 2010. "White Matter Integrity in Mild Cognitive Impairment: A Tract-based Spatial Statistics Study." *Neuroimage* 53(1):16–25.

13 Immunoscore, Circulating Tumor Cells, and Human-Derived Organoids as Potential Predictive Tools in Personalized Cancer Medicine

Agnieszka Pastuła

Introduction

Cancer is one of the leading causes of morbidity and mortality worldwide. Despite the enormous progress in cancer research during the last decade, as an extremely complex disease involving genetic and epigenetic changes in cancer cells, as well as alterations in the tumor microenvironment, the vast majority of cancers remain incurable. One of the main challenges in oncology lies in the different response to drugs in patients with the same type and grade of cancer. Among the factors which contribute to this phenomenon are genetic variability in the human population and tumor heterogeneity. One could even argue that every tumor is unique. This highlights the need for personalized approaches in cancer medicine. Accurate diagnostic and predictive tools are crucial. There is an urgent need for sensitive prognostic and predictive factors in order to identify patients who respond to a given therapy and patients who are at higher risk of tumor recurrence, as well as to stratify patients and monitor response to therapy.

From Tumor Microenvironment to Development of the Immunoscore

Currently the tumor staging system (TNM) is the main tool utilized as a prognostic predictor. The TNM staging system is based on the analysis of tumor burden (T), presence of cancer cells in lymph nodes (N), and distant metastasis (M), thus focusing mainly on tumor cells. Besides cancer cells themselves, there is a lot of evidence that non-genetic factors such as microenvironment also contribute to the phenotype of a malignant disease. Many authors have shown that the microenvironment plays a pivotal role in all steps during tumorigenesis: tumor initiation, tumor progression, and metastasis (Bhowmick et al. 2004; Joyce and Pollard 2009), and contributes to resistance to therapy (Tredan et al. 2007). The tumor microenvironment is composed of an extracellular matrix, soluble factors, and different types of stromal cells, such as endothelial cells, carcinoma associated fibroblasts, neurons, and immune cells (Liotta and Kohn 2001; Hanahan and Weinberg 2011; Quante et al. 2011).

More than a century ago, it was observed that the immune system is able to react against a tumor (Balkwill and Mantovani 2001). However, it is now known that immunity to a tumor is a double-edged sword: Depending on the type of immune cell or type of inflammatory reaction (acute vs. chronic; Th1 vs. Th2 response), the immune system might either eliminate tumor cells or promote tumor growth (O'Byrne and Dalgleish 2001; DeNardo and Coussens 2007). The interrelationship between the immune system and the cancer was described as cancer immunoediting, which is composed of three phases: immune surveillance, escape, and equilibrium (Dunn et al. 2004). During immune surveillance (elimination), cancer cells are destroyed by immune cells. If this clearance is successful, there is no clinically measurable disease. Should some cancer cells manage to avoid an immune cell attack (appearance of resistant clones), this is known as the escape phase. In this case, the surviving cancer cells divide and the tumor progresses. During the equilibrium phase, which is typically the phase when the cancer is diagnosed in the patients, immune cells are still capable of eliminating tumor cells, but are unable to completely eradicate the tumor cells.

From a clinical point of view it is important to decipher whether the presence of distinct immune cell populations at the tumor site could have an impact on the patient's outcome. In a study involving a cohort of 415 patients with colorectal cancer, it was shown that immune cell type, density, and location within the tumor could be a prognostic factor (Galon et al. 2006). The following markers of adaptive immune response were assessed: CD3 (total T cells), CD8 (cytotoxic T lymphocytes), and CD45RO (memory T cells). The study revealed that a high density of CD3+CD8+CD45RO+ T cells was associated with longer disease-free survival (DFS) and overall survival (OS). Moreover, the study showed that the type, density, and location of tumor infiltrating lymphocytes seemed to be a better prognostic factor than the TNM staging system. The infiltration of CD8+CD45RO+ T cells has been also shown to be associated with good prognosis in patients with different types of tumors, such as breast cancer, lung cancer, prostate cancer, and melanoma (Fridman et al. 2012). Besides the association of immune infiltrates with survival, tumor infiltrating lymphocytes were also shown to have the potential to predict response to chemotherapy. In the retrospective study, it was discovered that colorectal patients with strong tumor infiltration of lymphocytes had prolonged survival after surgery followed by treatment with 5-fluorouracil (Morris et al. 2008). However, the results will have to be confirmed in a prospective study. In the future, studies on patients with different types of cancer need to be performed in order to evaluate the value of tumor infiltrating lymphocytes as a tool to predict response to therapy. The above results have led to the development of a standardized scoring system to quantify the lymphocytes that infiltrate the tumor, known as immunoscore (Galon et al. 2014). The system relies on enumeration of CD3+CD8+CD45RO+ T cells at the tumor center and margins. The results are given as number of cells per mm^2. In order to validate the immunoscore, an international consortium with twenty-three centers involved was initiated in 2012. It is planned to test immunoscore on tumor samples from several thousand colorectal cancer patients. The study aims to assess the feasibility, reproducibility, and prognostic value of the scoring system. Moreover, the question of whether the immunoscore could be used to predict which patients are at high risk of disease recurrence will be evaluated. Should it prove successful, the immunoscore may have the potential to be used for the modernization of the TNM cancer staging system (Galon et al. 2012; Galon et al. 2014).

Prognostic and Predictive Value of Tumor Circulating Cells

More than 90% of cancer-related deaths are due to metastasis. During tumor progression, tumor cells start to detach from the primary tumor, invade local tissues, and then enter the blood circulation. The tumor cells that are present in the bloodstream are called circulating tumor cells (CTCs) (Racila et al. 1998), and are believed to be the direct precursors of the metastases (Hristozova et al. 2011; Baccelli et al. 2013). CTCs are extraordinarily rare. It is estimated that approximately one cell per 10^9 hematologic cells in the blood is a CTC (Nagrath et al. 2007). CTCs are characterized by the expression of EpCAM and cytokeratin 8/18/19, and they are negative for the leukocyte marker, CD45. So far, Cell Search® is the only system capable of isolating CTCs approved by the Food and Drug Administration (FDA). The isolation procedure is based on the immunomagnetic enrichment of EpCAM positive cells, which is followed by immunofluorescence for cytokeratin 8/18/19 and CD45 (Yu et al. 2011). In addition to the CellSearch® system, other technologies are under development. One example is a microfluidic system, which enables detection of CTCs in non-epithelial cancers such as melanoma, and seems to be more sensitive than CellSearch® (Ozkumur et al. 2013). Moreover, CTCs can be detected by MAINTRAC® (which utilizes fluorochrome tagged anti-EpCAM and a laser scanning cytometer) (Pachmann et al. 2008), EPISPOT (which relies on the short-term culturing of CTCs followed by identification of secreted soluble factors that are specific to epithelial cells), CTC-chip, and polymerase chain reaction (PCR) (Allan and Keeney 2010). CTCs have been found to be present in cancer patients with metastatic disease in the majority of types of solid tumors. As CTCs are potential mediators of cancer progression, it was hypothesized that they could carry prognostic and predictive information in patients with cancer. Many trials have addressed this issue.

In a prospective, multicenter study involving 177 patients with *metastatic breast cancer* (Cristofanilli et al. 2004), CTCs were measured before therapy and at the first follow-up after the initiation of therapy. In breast cancer, the increased level of CTCs is defined as five or more CTCs per 7.5 ml of peripheral blood. It was observed that patients with increased numbers of CTCs at baseline had significantly shorter median progression-free survival (PFS) and shorter OS. Moreover, enumeration of CTCs at the first follow-up, which was performed three to four weeks after the initiation of the new therapy, revealed that within a group of patients with unfavorable prognoses, there was a subgroup that benefited from therapy. To conclude, the study showed that CTC level before treatment is an independent predictor of PFS and OS in patients with metastatic breast cancer.

In another prospective and multicenter study, which involved 430 patients with *metastatic colorectal cancer* (Cohen et al. 2008), the prognostic and predictive value of CTCs was investigated. In colorectal cancer, the threshold of CTCs is three CTCs per 7.5 ml of peripheral blood. Patients with an increased CTC count at baseline had significantly shorter median PFS and OS. CTC enumeration was performed at different time points after the initiation of treatment, and revealed that PFS and OS were significantly shorter for the patients with increased CTC count. Moreover, a follow-up imaging study was performed. The analysis was carried out according to Response Evaluation Criteria in Solid Tumors (RECIST). The correlation between CTC number and OS was found, which proved that CTCs can provide additional prognostic information. The limitation of the study was that it involved a heterogeneous group of patients in terms of treatment.

The SWOG S0500 trial addressed whether changing the chemotherapy regimen in patients with persistently increased CTCs will improve OS (Smerage et al. 2014). The randomized study included 595 patients with *metastatic breast cancer*. The trial revealed that switching to alternative cytotoxic therapy did not improve the OS of patients. This study confirms the prognostic value of CTCs, and it indicates that resistance to chemotherapy could develop in patients in whom CTC levels do not decrease after first line chemotherapy. This would suggest the need for novel therapies. For this group of patients, CTCs could be used for molecular analysis and drug sensitivity testing, for example for targeted therapies.

In addition to the potential of CTCs as a predictor of survival, many studies looked at the question whether CTCs could act as predictors of therapy response. Various other trials (Pachmann et al. 2005; Camara et al. 2007; Liu et al. 2009; Thalgott et al. 2015) have provided evidence that the level of CTCs can predict survival and therapy response in patients with different cancer types. Collectively, these studies clearly showed that patients with decreased levels of CTCs after the initiation of therapy are likely to benefit from the treatment, whereas patients with increased or persistently high levels of CTCs are unlikely to have a favorable clinical outcome from the ongoing treatment and should be offered an alternative therapy. Several studies on patients with metastatic cancer demonstrated that assessment of the disease status by CTC enumeration is comparable (or even superior) to tumor response evaluation by traditional imaging techniques, and is characterized by high sensitivity and specificity, as well as significant predictive accuracy

(Budd et al. 2006; Liu et al. 2009; Thalgott et al. 2015). Incorporation of CTC counts as a surrogate marker to evaluate therapeutic benefit in patients with cancer can contribute to the earlier detection of chemoresistance (thus sparing patient toxicity) and earlier regimen switch to more effective therapy. In addition, application of CTCs to monitor response to therapy could limit radiology-based follow-up studies, and as a result reduce patients' exposure to radiation. Therefore, the quantification of CTCs should be considered for routine testing at clinics.

Concept of Applying Human-Derived Organoids to Predict Response to Anti-Cancer Drugs

Since 1951, when malignant cells were first derived from cervical cancer and placed in a dish (known as the HeLa cell line), it has been possible to culture human cancer cells outside the body (Scherer et al. 1953). Since that time, culturing human cancer cells has contributed enormously to drug discovery, overall progress in cancer research, and the improved understanding of malignant disease (Shoemaker 2006). The *Cancer Cell Line Encyclopedia* was created recently (Barretina et al. 2012) and includes genomic data of 947 human cancer cell lines, thus representing variable types of human cancers. In addition, the pharmacologic profile of twenty-four drugs for approximately 500 human cancer cell lines has been generated. Although this system revealed novel candidate biomarkers to predict clinical response, it reflects the cancer heterogeneity among the patients only to a limited extent. Therefore, culturing a patient's cells to predict response to drugs could be the best option in personalized medicine, especially for the patients with refractory disease (Crystal et al. 2014).

Organoids represent cell cultures that resemble an organ (Sato et al. 2009; Pastula and Quante 2014; Cao et al. 2015). As cells are embedded in matrigel, which provides a scaffold for cells, organoid culture is considered a type of three dimensional (3D) cell culture system. These are alleged to better mimic native microenvironmental conditions than traditional two-dimensional (2D) cell cultures. 3D cell culture models were shown to better preserve the features of a primary tumor as compared to 2D systems (Lee et al. 2013). Moreover, tumor cells cultured in a 3D system generally seem to be more resistant to chemotherapeutics (Fischbach et al. 2007). Importantly, in the 3D cell culture system, it is possible to combine tumor cells together with stromal cells, and the latter have been shown in experimental models to influence the response to anti-cancer therapeutics (Loeffler et al. 2006; Wang et al. 2009). Taken together, 3D cultures are currently considered better models for testing sensitivity to anti-cancer drugs than 2D cultures (Leung et al. 2015). Initially, organoid cultures were developed from the small intestine of the mouse, and intestinal organoid culture was described as mini-gut culture (Sato et al. 2009). Recently, organoid cultures have also been established from human tissue, for example: colon cancer, pancreatic cancer, prostate cancer, and Barrett's Esophagus epithelium (Sato et al. 2011; Gao et al. 2014; Boj et al. 2015; Calon et al. 2015). Currently, human-derived organoids are undergoing initial evaluations for their applicability for drug testing. The Living BioBank of human-derived organoids, a collection of organoids from different types of tumors and healthy tissues, has been created for this purpose (Hubrecht Organoid Technology 2014). Future studies will show whether organoids can be a suitable tool for predicting response to drugs in patients with cancer.

Alongside organoids derived from primary tumor tissue, it would be clinically relevant to apply this culture system to CTCs. As CTCs are believed to be metastasis founder cells, they are promising target cells for novel anti-cancer therapies. So far, culturing CTCs-derived organoids has only been reported for prostate cancer (Gao et al. 2014). However, the authors only describe the culture of CTCs from one patient, so further studies are needed to confirm the findings. In addition, the establishment of CTCs-derived organoids from other types of cancer is needed. So, in the future, CTC enumeration could be used not only as a predictor of patient survival, for patient stratification, and monitoring response to therapy, but also as a direct target for therapy, and a tool for predicting response to drugs *ex vivo* (drug sensitivity tests on CTCs-derived organoids). I predict that the establishment of human CTCs-derived organoids will be the next key step in cancer research, and will significantly contribute to personalized cancer medicine. The patient's own CTCs-derived organoids could be applied especially for those patients for whom fresh biopsy material from the primary tumor tissue is not available.

Summary and Future Directions

An increasing volume of data points to the enormous tumor heterogeneity among patients with malignant disease. There is a growing body of evidence that suggests that successful cancer treatment requires predictive and personalized approaches. Observations from tumor immunology have contributed to the findings that immune cell type, density, and location within the tumor could be a prognostic factor, and have thus led to the development of the immunoscore. Future studies should confirm the prognostic value of this scoring system and indicate whether the immunoscore can be used as a tool to predict response to therapy.

In addition to the immunoscore, enumeration of CTCs is another example of a promising personalized approach in cancer medicine. CTC levels were shown to predict survival in patients with different types of cancer, and, importantly, they can also be used to predict response to therapy. Thus, CTCs seem to have the potential to be accepted as outcome surrogates in the future. The advantages of the method include its non-invasive character and its better sensitivity as compared to imaging techniques. Moreover, CTCs can be measured repeatedly thus allowing real-time analysis of the disease status and response to treatment and early detection of minimal residual disease, which all can ultimately lead to the adjustment of the most effective therapy to the particular patient, longer survival, and higher success rates in cancer medicine. Furthermore, incorporation of CTC enumeration into the clinic could bring potential advantages to patients, such as avoiding unnecessary toxicity, earlier switching to effective therapy, and reduced exposure to radiation. Besides this, the implementation of CTC counts into the clinical routine could reduce the risk of complications associated with taking biopsies, for example in patients with lung lesions, or patients with prostate cancer who are under active surveillance. However, large multicenter studies should be performed in order to confirm the clinical utility of CTC enumeration.

In the future, methods for the detection of CTCs should be improved, as the current markers do not allow identification of CTCs that are undergoing epithelial-mesenchymal transition (EMT), which is characterized by downregulation of epithelial markers and upregulation of mesenchymal markers. CTCs with EMT characteristics may prove to be the most resistant to therapy and the most aggressive.

Another important aspect which remains to be investigated is the molecular analysis of CTCs, and culturing the patient's CTCs for drug sensitivity tests. This could be performed by culturing isolated CTCs in the form of organoids. Currently, human organoids derived from primary tumors are tested for their potential to predict response to therapy. The establishment of the organoid culture system from liquid biopsy would appear to be the next crucial step. The patient's own CTCs cultured in a 3D culture system could be tested for drug sensitivity, thus enabling personalized and specific targeting of key players in metastatic disease—metastasis seeding cells.

Taken together, the implementation of immunoscore, quantification of CTC levels, and culturing of patients' own tumor cells in the form of organoids, could extend the TNM staging system, and lead to improved therapy guidance as well as to more personalized and more predictive approaches in cancer medicine. Although the bulk of validating work is still ahead, there is a tangible promise of increasing patients' life quality and reducing cancer-associated mortalities.

References

Allan, A. L., and M. Keeney. 2010. "Circulating Tumor Cell Analysis: Technical and Statistical Considerations for Application to the Clinic." *Journal of Oncology*: 426218–27

Baccelli, I., Schneeweiss, A., Riethdorf, S., Stenzinger, A., Schillert, A., Vogel, V., Klein, C., Saini, M., Bauerle, T., Wallwiener, M., Holland-Letz, T., Hofner, T., Sprick, M., Scharpff, M., Marme, F., Sinn, H. P., Pantel, K., Weichert, W. and Trumpp, A. 2013. "Identification of a Population of Blood Circulating Tumor Cells from Breast Cancer Patients that Initiates Metastasis in a Xenograft Assay." *Nature Biotechnology* 31(6):539–44.

Balkwill, F., and A. Mantovani. 2001. "Inflammation and Cancer: Back to Virchow?" *The Lancet* 357(9255):539–45.

Barretina, J., Caponigro, G., Stransky, N., Venkatesan, K., Margolin, A. A., Kim, S., Wilson, C. J., Lehar, J., Kryukov, G. V., Sonkin, D., Reddy, A., Liu, M., Murray, L., Berger, M. F., Monahan, J. E., Morais, P., Meltzer, J., Korejwa, A., Jane-Valbuena, J., Mapa, F. A., Thibault, J., Bric-Furlong, E., Raman, P., Shipway, A., Engels, I. H., Cheng, J., Yu, G. K., Yu, J., Aspesi, P., Jr., de Silva, M., Jagtap, K., Jones, M. D., Wang, L., Hatton, C., Palescandolo, E., Gupta, S., Mahan, S., Sougnez, C., Onofrio, R. C., Liefeld, T., MacConaill, L., Winckler, W., Reich, M., Li, N., Mesirov, J. P., Gabriel, S. B., Getz, G., Ardlie, K., Chan, V., Myer, V. E., Weber, B. L., Porter, J., Warmuth, M., Finan, P., Harris, J. L., Meyerson, M., Golub, T. R., Morrissey, M. P., Sellers, W. R., Schlegel, R. and Garraway, L. A. 2012. "The Cancer Cell Line Encyclopedia Enables Predictive Modelling of Anticancer Drug Sensitivity." *Nature* 483(7391):603–7.

Bhowmick, N. A., Neilson, E. G., and H. L. Moses. 2004. "Stromal Fibroblasts in Cancer Initiation and Progression." *Nature* 432(7015):332–7.

Boj, S. F., Hwang, C. I., Baker, L. A., Chio, II, Engle, D. D., Corbo, V., Jager, M., Ponz-Sarvise, M., Tiriac, H., Spector, M. S., Gracanin, A., Oni, T., Yu, K. H., van Boxtel, R., Huch, M., Rivera, K. D., Wilson, J. P., Feigin, M. E., Ohlund, D., Handly-Santana, A., Ardito-Abraham, C. M., Ludwig, M., Elyada, E., Alagesan, B., Biffi, G., Yordanov, G. N., Delcuze, B., Creighton, B., Wright, K., Park, Y., Morsink, F. H., Molenaar, I. Q., Borel Rinkes, I. H., Cuppen, E., Hao, Y., Jin, Y., Nijman, I. J., Iacobuzio-Donahue, C., Leach, S. D., Pappin, D. J., Hammell, M., Klimstra, D. S., Basturk, O., Hruban, R. H., Offerhaus, G. J., Vries, R. G., Clevers, H. and Tuveson, D. A. 2015. "Organoid Models of Human and Mouse Ductal Pancreatic Cancer." *Cell* 160(1–2):324–38.

Budd, G. T., Cristofanilli, M., Ellis, M.J., Stopeck, A., Borden, E., Miller, M. C., Matera, J., Repollet, M., Doyle, G. V., Terstappen, L. W., and D. F. Hayes. 2006. "Circulating Tumor Cells Versus Imaging—Predicting Overall Survival in Metastatic Breast Cancer." *Clinical Cancer Research* 12(21):6403–9.

Calon, A., Lonardo, E., Berenguer-Llergo, A., Espinet, E., Hernando-Momblona, X., Iglesias, M., Sevillano, M., Palomo-Ponce, S., Tauriello, D. V., Byrom, D., Cortina, C., Morral, C., Barcelo, C., Tosi, S., Riera, A., Attolini, C. S., Rossell, D., Sancho, E. and Batlle, E. 2015. "Stromal Gene Expression Defines Poor-Prognosis Subtypes in Colorectal Cancer." *Nature Genetics* 47(4):320–9.

Camara, O., Rengsberger, M., Egbe, A., Koch, A., Gajda, M., Hammer, U., Jorke, C., Rabenstein, C., Untch, M., and K. Pachmann. 2007. "The Relevance of Circulating Epithelial Tumor Cells (CETC) for Therapy Monitoring during Neoadjuvant (Primary Systemic) Chemotherapy in Breast Cancer." *Annals of Oncology* 18(9):1484–92.

Cao, L., Kuratnik, A., Xu, W., Gibson, J. D., Kolling, F. t., Falcone, E. R., Ammar, M., Van Heyst, M. D., Wright, D. L., Nelson, C. E. and Giardina, C. 2015. "Development of Intestinal Organoids as Tissue Surrogates: Cell Composition and the Epigenetic Control of Differentiation." *Molecular Carcinogenesis* 54(3):189–202.

Cohen, S. J., Punt, C. J., Iannotti, N., Saidman, B. H., Sabbath, K. D., Gabrail, N. Y., Picus, J., et al. 2008. "Relationship of Circulating Tumor Cells to Tumor Response, ProgressionFree Survival, and Overall Survival in Patients with Metastatic Colorectal Cancer." *Journal of Clinical Oncology* 26(19):3213–21.

Cristofanilli, M., Budd, G. T., Ellis, M. J., Stopeck, A., Matera, J., Miller, M. C., Reuben, J. M., Doyle, G. V., Allard, W. J., Terstappen, L. W. and Hayes, D. F. 2004. "Circulating Tumor Cells, Disease Progression, and Survival in Metastatic Breast Cancer." *The New England Journal of Medicine* 351(8):781–91.

Crystal, A. S., Shaw, A. T., Sequist, L. V., Friboulet, L., Niederst, M. J., Lockerman, E. L., Frias, R. L., Gainor, J. F., Amzallag, A., Greninger, P., Lee, D., Kalsy, A., Gomez-Caraballo, M., Elamine, L., Howe, E., Hur, W., Lifshits, E., Robinson, H. E., Katayama, R., Faber, A. C., Awad, M. M., Ramaswamy, S., Mino-Kenudson, M., Iafrate, A. J., Benes, C. H. and Engelman, J. A. 2014. "Patient-Derived Models of Acquired Resistance Can Identify Effective Drug Combinations for Cancer." *Science* 346(6216):1480–6.

DeNardo, D. G., and L. M. Coussens. 2007. "Inflammation and Breast Cancer. Balancing Immune Response: Crosstalk between Adaptive and Innate Immune Cells during Breast Cancer Progression." *Breast Cancer Research* 9(4):212.

Dunn, G. P., Old, L. J., and R. D. Schreiber. 2004. 'The Three Es of Cancer Immunoediting." *Annual Reviews of Immunology* 22:329–60.

Fischbach, C., Chen, R., Matsumoto, T., Schmelzle, T., Brugge, J. S., Polverini, P. J., and D. J. Mooney. 2007. "Engineering Tumors with 3D Scaffolds." *Nature Methods* 4(10):855–60.

Fridman, W. H., Pages, F., Sautes-Fridman, C., and J. Galon. 2012. "The Immune Contexture in Human Tumours: Impact on Clinical Outcome." *Nature Reviews Cancer* 12(4):298–306.

Galon, J., Costes, A., Sanchez-Cabo, F., Kirilovsky, A., Mlecnik, B., Lagorce-Pages, C., Tosolini, M., Camus, M., Berger, A., Wind, P., Zinzindohoue, F., Bruneval, P., Cugnenc, P. H., Trajanoski, Z., Fridman, W. H. and Pages, F. 2006. "Type, Density, and Location of Immune Cells within Human Colorectal Tumors Predict Clinical Outcome." *Science* 313(5795): 1960–4.

Galon, J., Mlecnik, B., Bindea, G., Angell, H. K., Berger, A., Lagorce, C., Lugli, A., et al. 2014. "Towards the Introduction of the 'Immunoscore' in the Classification of Malignant Tumours." *The Journal of Pathology* 232(2):199–209.

Galon, J., Pages, F., Marincola, F. M., Angell, H. K., Thurin, M., Lugli, A., Zlobec, I., et al. 2012. "Cancer Classification using the Immunoscore: A Worldwide Task Force." *Journal of Translational Medicine* 10:205.

Gao, D., Vela, I., Sboner, A., Iaquinta, P. J., Karthaus, W. R., Gopalan, A., Dowling, C., Wanjala, J. N., Undvall, E. A., Arora, V. K., Wongvipat, J., Kossai, M., Ramazanoglu, S., Barboza, L. P., Di, W., Cao, Z., Zhang, Q. F., Sirota, I., Ran, L., MacDonald, T. Y., Beltran, H., Mosquera, J. M., Touijer, K. A., Scardino, P. T., Laudone, V. P., Curtis, K. R., Rathkopf, D. E., Morris, M. J., Danila, D. C., Slovin, S. F., Solomon, S. B., Eastham, J. A., Chi, P., Carver, B., Rubin, M. A., Scher, H. I., 2014. Clevers, H., Sawyers, C. L. and Chen, Y. "Organoid Cultures Derived from Patients with Advanced Prostate Cancer." *Cell* 159(1):176–87.

Hanahan, D., and R. A. Weinberg. 2011. "Hallmarks of Cancer: The Next Generation." *Cell* 144(5):646–74.

Hristozova, T., Konschak, R., Stromberger, C., Fusi, A., Liu, Z., Weichert, W., Stenzinger, A., Budach, V., Keilholz, U., and I. Tinhofer. 2011. "The Presence of Circulating Tumor Cells (CTCs) Correlates with Lymph Node Metastasis in Nonresectable Squamous Cell Carcinoma of the Head and Neck Region (SCCHN)." *Annals of Oncology* 22(8):1878–85.

Hubrecht Organoid Technology. 2014. Available at http://hub4organoids.eu [accessed March 14, 2016].

Joyce, J.A., and J.W. Pollard. 2009. "Microenvironmental Regulation of Metastasis." *Nature Reviews Cancer* 9(4):239–52.

Lee, J.M., Mhawech-Fauceglia, P., Lee, N., Parsanian, L.C., Lin, Y.G., Gayther, S.A., and K. Lawrenson. 2013. "A Three-Dimensional Microenvironment Alters Protein Expression and Chemosensitivity of Epithelial Ovarian Cancer Cells in Vitro." *Laboratory Investigation* 93(5):528–42.

Leung, B.M., Moraes, C., Cavnar, S.P., Luker, K.E., Luker, G.D., and S. Takayama. 2015. "Microscale 3D Collagen Cell Culture Assays in Conventional Flat-Bottom 384-Well Plates." *Journal of Laboratory Automation* 20(2):138–45.

Liotta, L.A., and E.C. Kohn. 2001. "The Microenvironment of the Tumour-Host Interface." *Nature* 411(6835):375–9.

Liu, M. C., Shields, P. G., Warren, R. D., Cohen, P., Wilkinson, M., Ottaviano, Y. L., Rao, S. B., Eng-Wong, J., Seillier-Moiseiwitsch, F., Noone, A. M. and Isaacs, C. 2009. "Circulating Tumor Cells: A Useful Predictor of Treatment Efficacy in Metastatic Breast Cancer." *Journal of Clinical Oncology* 27(31):5153–9.

Loeffler, M., Kruger, J.A., Niethammer, A.G., and R.A. Reisfeld. 2006. "Targeting Tumor-Associated Fibroblasts Improves Cancer Chemotherapy by Increasing Intratumoral Drug Uptake." *Journal of Clinical Investigation* 116(7):1955–62.

Morris, M., Platell, C., and B. Iacopetta. 2008. "Tumor-Infiltrating Lymphocytes and Perforation in Colon Cancer Predict Positive Response to 5-Fluorouracil Chemotherapy." *Clinical Cancer Research* 14(5):1413–7.

Nagrath, S., Sequist, L. V., Maheswaran, S., Bell, D. W., Irimia, D., Ulkus, L., Smith, M. R., Kwak, E. L., Digumarthy, S., Muzikansky, A., Ryan, P., Balis, U. J., Tompkins, R. G., Haber, D. A. and Toner, M. 2007. "Isolation of Rare Circulating Tumour Cells in Cancer Patients by Microchip Technology." *Nature* 450(7173):1235–9.

O'Byrne, K.J., and A.G. Dalgleish. 2001. Chronic Immune Activation and Inflammation as the Cause of Malignancy." *British Journal of Cancer* 85(4):473–83.

Ozkumur, E., Shah, A. M., Ciciliano, J. C., Emmink, B. L., Miyamoto, D. T., Brachtel, E., Yu, M., Chen, P. I., Morgan, B., Trautwein, J., Kimura, A., Sengupta, S., Stott, S. L., Karabacak, N. M., Barber, T. A., Walsh, J. R., Smith, K., Spuhler, P. S., Sullivan, J. P., Lee, R. J., Ting, D. T., Luo, X., Shaw, A. T., Bardia, A., Sequist, L. V., Louis, D. N., Maheswaran, S., Kapur, R., Haber, D. A. and Toner, M. 2013. "Inertial Focusing for Tumor Antigen-Dependent and -Independent Sorting of Rare Circulating Tumor Cells." *Science Translational Medicine* 5(179):179ra47.

Pachmann, K., Camara, O., Kavallaris, A., Krauspe, S., Malarski, N., Gajda, M., Kroll, T., Jorke, C., Hammer, U., Altendorf-Hofmann, A., Rabenstein, C., Pachmann, U., Runnebaum, I. and Hoffken, K. 2008. "Monitoring the Response of Circulating Epithelial Tumor Cells to Adjuvant Chemotherapy in Breast Cancer Allows Detection of Patients at Risk of Early Relapse." *Journal of Clinical Oncology* 26(8):1208–15.

Pachmann, K., Camara, O., Kavallaris, A., Schneider, U., Schunemann, S., and K. Hoffken. 2005. "Quantification of the Response of Circulating Epithelial Cells to Neodadjuvant Treatment for Breast Cancer: A New Tool for Therapy Monitoring." *Breast Cancer Research* 7(6):R975–9.

Pastuła, A. and Quante, M. (2014). Isolation and 3-dimensional Culture of Primary Murine Intestinal Epithelial Cells. *Bio-protocol* 4(10): e1125.

Quante, M., Tu, S. P., Tomita, H., Gonda, T., Wang, S. S., Takashi, S., Baik, G. H., Shibata, W., Diprete, B., Betz, K. S., Friedman, R., Varro, A., Tycko, B. and Wang, T. C. 2011. "Bone Marrow-Derived Myofibroblasts Contribute to the Mesenchymal Stem Cell Niche and Promote Tumor Growth." *Cancer Cell* 19(2):257–72.

Racila, E., Euhus, D., Weiss, A.J., Rao, C., McConnell, J., Terstappen, L.W., and J.W. Uhr. 1998. "Detection and Characterization of Carcinoma Cells in the Blood." *Proceedings of the National Academy of Sciences USA* 95(8):4589–94.

Sato, T., Stange, D. E., Ferrante, M., Vries, R. G., Van Es, J. H., Van den Brink, S., Van Houdt, W. J., Pronk, A., Van Gorp, J., Siersema, P. D. and Clevers, H. 2011. "Long-Term Expansion of Epithelial Organoids from Human Colon, Adenoma, Adenocarcinoma, and Barrett's Epithelium." *Gastroenterology* 141(5):1762–72.

Sato, T., Vries, R. G., Snippert, H. J., van de Wetering, M., Barker, N., Stange, D. E., van Es, J. H., Abo, A., Kujala, P., Peters, P. J. and Clevers, H. 2009. "Single Lgr5 Stem Cells Build Crypt-Villus Structures in Vitro without a Mesenchymal Niche." *Nature* 459(7244):262–5.

Scherer, W.F., Syverton, J.T., and G.O. Gey. 1953. "Studies on the Propagation in Vitro of Poliomyelitis Viruses: Viral Multiplication in a Stable Strain of Human Malignant Epithelial Cells (Strain HeLa) Derived from an Epidermoid Carcinoma of the Cervix." *Journal of Experimental Medicine* 97(5):695–710.

Shoemaker, R.H. 2006. "The NCI60 Human Tumour Cell Line Anticancer Drug Screen." *Nature Reviews Cancer* 6(10):813–23.

Smerage, J. B., Barlow, W. E., Hortobagyi, G. N., Winer, E. P., Leyland-Jones, B., Srkalovic, G., Tejwani, S., Schott, A. F., O'Rourke, M. A., Lew, D. L., Doyle, G. V., Gralow, J. R., Livingston, R. B. and Hayes, D. F. 2014. "Circulating Tumor Cells and Response to Chemotherapy in Metastatic Breast Cancer: SWOG S0500." *Journal of Clinical Oncology* 32(31):3483–9.

Thalgott, M., Heck, M. M., Eiber, M., Souvatzoglou, M., Hatzichristodoulou, G., Kehl, V., Krause, B. J., Rack, B., Retz, M., Gschwend, J. E., Andergassen, U. and Nawroth, R. 2015. "Circulating Tumor Cells Versus Objective Response Assessment Predicting Survival in Metastatic Castration-Resistant Prostate Cancer Patients Treated with Docetaxel Chemotherapy." *Journal of Cancer Research and Clinical Oncology* 141(8):1457–64.

Tredan, O., Galmarini, C.M., Patel, K., and I.F. Tannock. 2007. "Drug Resistance and the Solid Tumor Microenvironment." *Journal of National Cancer Institute* 99(19):1441–54.

Wang, W., Li, Q., Yamada, T., Matsumoto, K., Matsumoto, I., Oda, M., Watanabe, G., Kayano, Y., Nishioka, Y., Sone, S. and Yano, S. 2009. "Crosstalk to Stromal Fibroblasts Induces Resistance of Lung Cancer to Epidermal Growth Factor Receptor Tyrosine Kinase Inhibitors." *Clinical Cancer Research* 15(21):6630–8.

Yu, M., Stott, S., Toner, M., Maheswaran, S., and D.A. Haber. 2011. "Circulating Tumor Cells: Approaches to Isolation and Characterization." *Journal of Cell Biology* 192(3):373–82.

Selected Bibliography

Altman, D.G., and P. Royston. 2000. "What Do We Mean by Validating a Prognostic Model?" *Statistics in Medicine* 19(4):453–73.

Beauchamp, T.L., and J. F. Childress. 2009. *Principles of Biomedical Ethics*, 6th ed. Oxford: Oxford University Press.

Beck, U. 1992. *Risk Society*. London: Sage.

British Medical Association. 1998. *Human Genetics: Choice and Responsibility*. Oxford: Oxford University Press.

Caplan, A.L., McCartney, J.J., and D.A. Sisti, eds. 2004. *Health, Disease and Illness: Concepts in Medicine*. Washington: Georgetown University Press.

Childs, B. 1999. *Genetic Medicine: A Logic of Disease*. Baltimore/London: John Hopkins University Press.

Clark, W.R. 1997. *The New Healers: The Promise and Problems of Molecular Medicine in the Twenty-First Century*. New York/Oxford: Oxford University Press.

Collins, F.C. 2011. *The Language of Life: DNA and the Revolution in Personalized Medicine*. New York: Harper Perennial.

Dabrock, P., Braun, M., and J. Ried, eds. 2012. *Individualized Medicine between Hype and Hope: Exploring Ethical and Societal Challenges for Healthcare*. Vienna: LIT.

Fischer, T., Langanke, M., Marschall, P., and S. Michl, eds. 2015. *Individualized Medicine: Ethical, Economical and Historical Perspectives*. Cham, s.l.: Springer International.

Gigerenzer, G. 2003. *Reckoning with Risk: Learning to Live with Uncertainty*. London: Penguin.

Gordon, J., Macnaughton, J., and C.E. Rudebeck, eds. 2014. *Prognosis*. London, New York: Radcliffe Publishing. Medical Humanities Companion, Vol. 4.

Grech, G., and I. Grossman, eds. 2015. *Preventive and Predictive Genetics: Towards Personalised Medicine*. Cham, s.l.: Springer International.

Grove, W.M., Zald, D.H., Lebow, B.S., Snitz, B.E., and C. Nelson. 2000. "Clinical versus Mechanical Prediction: A Meta-Analysis." *Psychological Assessment* 12(1):19–30.

Habermas, J. 2003. *The Future of Human Nature*. Cambridge: Polity.

Hedgecoe, A. 2004. *The Politics of Personalised Medicine: Pharmacogenetics in the Clinic*. Cambridge/New York: Cambridge University Press. Cambridge Studies in Society and the Life Sciences.

Human Genetics Commission. 2006. *Making Babies: Reproductive Decisions and Genetic Testing.* Available at http://centronuclear.org.uk/theinformationpoint/documents/research/articles/making_babies.pdf [accessed March 16, 2016].

Juengst, E.T., Flatt, M.A., and R.A. Settersten. 2012. "Personalized Genomic Medicine and the Rhetoric of Empowerment." *Hastings Center Report* 42(5):34–40.

Karafyllis, N.C., and J.C. Schmidt, eds. 2002. *Zugänge zur Rationalität der Zukunft: M-&-P-Schriftenreihe für Wissenschaft und Forschung Philosophie.* Stuttgart: Metzler.

Kollek, R., and T. Lemke. 2008. *Der medizinische Blick in die Zukunft: Gesellschaftliche Implikatioenen prädiktiver Gentests.* New York: Frankfurt am Main.

Konrad, M. 2005. *Narrating the New Predictive Genetics: Ethics, Ethnography, and Science.* Cambridge, New York: Cambridge University Press. Cambridge Studies in Society and the Life Sciences.

Lenz, H.-J., ed. 2013. *Biomarkers in Oncology: Prediction and Prognosis.* New York: Springer.

Nuffield Council on Bioethics. 1993. *Nuffield Report on Genetic Screening: Ethical Issues.* London: Nuffield Council on Bioethics. Available at http://nuffieldbioethics.org/wp-content/uploads/2014/07/Genetic_screening_report.pdf [accessed March 10, 2016].

Nuffield Council on Bioethics. 2003. *Pharmacogenetics. Ethical Issues.* London: Nuffield Council on Bioethics. Available at http://nuffieldbioethics.org/wp-content/uploads/2014/07/Pharmacogenetics-Report.pdf [accessed March 10, 2016].

Pfleiderer, G., Battegay, M., and K. Lindpaintner, eds. 2012. *Knowing One's Medical Fate in Advance: Challenges for Diagnosis and Treatment, Philosophy, Ethics and Religion.* Basel/Freiburg/Paris/London/New York/New Delhi/Bangkok/Beijing/Tokyo/Kuala Lumpur/Singapore/Sydney: Karger.

Presidential Commission for the Study of Bioethical Issues. 2013. *Anticipate and Communicate: Ethical Management of Incidental and Secondary Findings in the Clinical, Research, and Direct-to-Consumer Contexts.* Available at http://bioethics.gov/sites/default/files/FINALAnticipateCommunicate_PCSBI_0.pdf [accessed March 10, 2016].

Propping, P., Aretz, S., Schumacher, J., Taupitz, J., Guttmann, J., and B. Heinrichs. 2006. *Prädiktive genetische Testverfahren: Naturwissenschaftliche, rechtliche und ethische Aspekte.* Freiburg/München: Karl Alber.

Rawls, J. 1971. *A Theory of Justice.* Cambridge: Harvard University Press.

Rehmann-Sutter, C., ed. 2009. *Disclosure Dilemmas: Ethics of Genetic Prognosis after the "Right to Know/Not to Know" Debate.* Farnham: Ashgate.

Rothstein, M.A., ed. 2003. *Pharmacogenomics: Social, Ethical, and Clinical Dimensions.* Hoboken: John Wiley & Sons.

Sandel, M.J. 2007. *The Case against Perfection: Ethics in the Age of Genetic Engineering.* Cambridge: Harvard University Press.

Vollmann, J., Schildmann, J., Wäscher, S., and V. Sandow, eds. 2015. *The Ethics of Personalised Medicine: Critical Perspectives.* Surrey: Ashgate.

Wehling, P., ed. 2015. *The Public Shaping of Medical Research: Patient Associations, Health Movements and Biomedicine.* London: Routledge. Routledge Studies in the Sociology of Health and Illness.

Contributors

Giovanni Boniolo, Dipartimento di Scienze Biomediche e Chirurgico Specialistiche, Università di Ferrara, Italy and Institute for Advanced Study, Technical University of Munich, Germany. He works on the philosophical foundations and the ethical implications of research biomedicine and clinical practice.

Florian Braune is a researcher at the Department for the History, Philosophy and Ethics of Medicine, University of Duesseldorf, Germany. His main research interests are political philosophy with regard to medicine and questions of autonomy.

Peter Conrad is the Harry Coplan Professor of Social Sciences in the Department of Sociology at Brandeis University, MA, USA. He has published several books on medicalization including *The Medicalization of Society* (2007) and the forthcoming co-edited *Global Perspectives on ADHD*.

Heiner Fangerau is director of the Department for the History, Philosophy and Ethics of Medicine, University of Duesseldorf, Germany. His main research topics are history of biomedicine, research ethics, and evolution of diagnostics.

Giulia Ferretti is a PhD student of the Dipartimento di Scienze della Salute, University of Milano in Italy. Her research focuses on the ethical issues related to cancer screening programs.

Steffen Flessa is head of the department of health care management of the faculty of law and economics of the University of Greifswald. His main research interests are quantitative methods of health care management, international health economics, and non-profit organizations. He is also a member of the faculty of medicine of the University of Greifswald and a frequent consultant for development agencies.

Mariacarla Gadebusch Bondio is full professor for History, Theory, and Ethics of Medicine. She is head of the Institute for History of Medicine at Rheinische Friedrich-Wilhelms-Universität Bonn and President of the Professional Association of Medical Historians in the German speaking world. Her research lies at the intersection of medicine and philosophy and focusses on the history of medical ethics and premodern medical culture.

Francesco Garaci is an associate professor in Radiology, head of the Operating Unit of Functional Studies at the Department of Diagnostic Imaging, Molecular Imaging, Interventional Radiology and Radiotherapy of the University Hospital of Rome "Tor Vergata." His clinical expertise is in the area of neuroradiology. His current research focuses on neuroimaging biomarkers for neurodegenerative diseases.

John-Stewart Gordon is full professor and head of the Research Cluster for Applied Ethics (RCAE) at Vytautas Magnus University in Kaunas, Lithuania. Furthermore, he is a member of the board of *Bioethics* since 2007, general editor and board member of the book series "Philosophy and Human Rights" at Brill, and has been area-editor and board member of the *Internet Encyclopedia of Philosophy* (2007–2014). He has written and edited several books in the context of practical philosophy and published peer-reviewed articles and special issues at international leading journals and encyclopedias.

Harald Hampel is full professor and AXA Research Fund and UPMC Chair at the Sorbonne Universities, Université Pierre et Marie Curie. He is Scientific Director of the Institute of Memory and Alzheimer's disease at the Department of Neurology, Pitié Salpêtrière University Hospital, Paris, France. His interests embrace the neurologic and psychiatric disorders, precision medicine, systems biology and neurophysiology, biological markers, neuroimaging and genetics of neurodegenerative diseases, as well as clinical trials and drug development for neurodegenerative diseases.

Elke Holinski-Feder is a clinical geneticist and co-founder and director of the MGZ–Medical Genetics Center in Munich, Germany. She is Professor of Medicine at Ludwig Maximilian University Medical Clinic. Her research focuses, among other areas, on hereditary gastrointestinal cancer syndromes.

Christian Lenk is managing director of the Research Ethics Committee and associate professor for medical ethics at the University of Ulm, Germany. His main research topics are ethics of biobanking, research ethics, and questions of the quality of life.

Stefan F. Lichtenthaler holds the chair for neuroproteomics (biochemistry) at the German Center for Neurodegenerative Diseases (DZNE) and the

Technical University of Munich (TUM), Germany. He is also a member of the Institute for Advanced Study of TUM and of the Munich Exellence Center for Systems Neurology (SyNergy). His research interest is in the molecular causes of neurodegeneration.

Alma Linkeviciute is a doctoral student at the European Institute of Oncology (IEO) and Dipartimento di Scienze della Salute, University of Milano, Milan, Italy. Her research interests lie in finding practically applicable solutions to ethical problems arising in oncology.

Simone Lista is a postdoctoral fellow at Pierre and Marie Curie University, Sorbonne Universities, Paris, France. After completing his combined BSc–MSc degree in Biology and his PhD degree in Clinical Biochemistry and Laboratory Medicine at the University of Pavia, he attended the European School for Scientific and Regulatory Assessment of New Medicines at the University of Rome "Tor Vergata" and moved to the Department of Psychiatry at Goethe-University of Frankfurt am Main. His current research focuses on the development and validation of multimodal biological markers for Alzheimer's disease.

Konrad Ott is director of the Department of Philosophy at Kiel University and Professor for Philosophy and Ethics of the Environment. The focus of his research lies on discourse ethics, bioethics, sustainability, and environmental ethics. At Greifswald University, he researched the idea and concepts of individualized medicine within the Gani_Med project. The article is an outcome of this research.

Agnieszka Pastuła is a researcher at the Department of Medicine II, Klinikum rechts der Isar, Technical University of Munich, Munich, Germany. She is investigating stromal-epithelial interactions during homeostasis and tumorigenesis using different three-dimensional cell culture models.

Francesco Spöring was a researcher at the Institute for History and Ethics of Medicine, Technical University of Munich. His current research focuses on the history of psychopharmacology.

Verena Steinke-Lange is a clinical geneticist at the MGZ–Medical Genetics Center in Munich, Germany. Her research focuses on hereditary colorectal cancer.

Nicola Toschi is an associate professor in Medical Physics at the Department of Biomedicine and Prevention, University of Rome "Tor Vergata" and a Visiting Assistant Professor and assistant in Neurology at the Athinoula A. Martinos Center for Biomedical Imaging (Harvard Medical School, Boston, USA). His research is focused on MR physics and developing image processing and modeling techniques able to extract quantitative biomarkers of investigative, diagnostic, and prognostic value in a clinical context.

Miranda Waggoner is assistant professor of Sociology at Florida State University, FL, USA. She studies the social and cultural dimensions of medicine, science, and public health, with an emphasis on the politics of gender and reproduction.

Xian-Ning is the secretary-general of the China Healthy Birth Science Association. He is vice-chair of the Department of Cell Biology and Medical Genetics, Zhejiang University School of Medicine at Hangzhou. He has devoted much of his life to the molecular and prenatal diagnosis of genetic disorders.

Ji Zuo is the chair of the China Healthy Birth Science Association. He heads the Department of Cellular and Genetic Medicine at Fudan University Shanghai Medical School. He has devoted himself to *yousheng* in China for more than twenty-eight years.

Index